FOOD, INC. 2

PARTICIPANT

FOOD, INC. 2

Inside the Quest for
a Better Future for Food

EDITED BY

KARL WEBER

PUBLICAFFAIRS

New York

PublicAffairs
Hachette Book Group
1290 Avenue of the Americas, New York, NY 10104
www.publicaffairsbooks.com
@Public Affairs

Printed in the United States of America

First Edition: October 2023

Published by PublicAffairs, an imprint of Perseus Books, LLC, a subsidiary of Hachette Book Group, Inc. The PublicAffairs name and logo is a registered trademark of the Hachette Book Group.

The Hachette Speakers Bureau provides a wide range of authors for speaking events. To find out more, go tohachettespeakersbureau.com or email HachetteSpeakers@hbgusa.com.

PublicAffairs books may be purchased in bulk for business, educational, or promotional use. For more information, please contact your local bookseller or the Hachette Book Group Special Markets Department at special.markets@hbgusa.com.

The publisher is not responsible for websites (or their content) that are not owned by the publisher.

Library of Congress Control Number: 2023946869

ISBNs: 9781541703575 (trade paperback), 9781541703582 (ebook)

LSC-C

Printing 1, 2023

CONTENTS

PREFACE

In 2009, I was lucky enough to be asked to edit the companion book to a remarkable new movie. *Food, Inc.* was a groundbreaking documentary film directed by Robby Kenner that transformed the way millions of people think about the food we eat. Narrated by journalists Michael Pollan and Eric Schlosser, it exposed the realities of corporate agriculture in the United States and, increasingly, around the world, showing its impact on the lives of workers, on the economies of rural communities, on the health of consumers, and on the environment of planet Earth. Critics responded to the film with phrases like "frankly riveting," "essential viewing," and "one of the year's most important films." *Food, Inc.* was nominated for awards in the documentary film category at the Eighty-Second Academy Awards and at the Twenty-Fifth Independent Spirit Awards, and it garnered the kind of audience attention few documentaries received.

I say I was lucky to be the editor of the companion book because that book became, for me, not just another interesting project but a remarkable publishing journey.

The film producers at Participant and River Road Entertainment, the publishing team led by Clive Priddle at PublicAffairs, and I were all determined to make the companion book more than just a collection of pictures and quotations from the movie. Instead, it would expand the important story told in the film into new areas, offering readers fresh insights into the food system and the ways it needs to change. We were able to carry out that ambitious program. *Food, Inc.* (the book) contained chapters by a who's who of experts and luminaries from the world of food reform, including both Schlosser and Pollan as well as Marion Nestle, Anna Lappé, Muhammad Yunus, and many more, and it delved deeply into issues ranging from the global war on hunger and the benefits of organic foods to the exploitation of farmworkers and the impact of factory farming on the Earth's environment.

The book became a major event in its own right. It appeared on the *New York Times* bestseller list, was translated into dozens of languages, and reached an enormous audience that complemented the one reached by the movie, including many thousands of high school and college students who experienced the book as supplementary reading for classes and discussion groups. It's widely read and discussed to this day.

Now the release of the much-anticipated sequel to *Food, Inc.* (the film) creates the opportunity to release a new companion book that will address the remarkable developments in the world of food over the past decade and a half—as well as the continuing challenges faced by those working to reform the system. The book you hold in your hand features a new roster of experts and activists who are tackling such issues as the climate impacts of industrialized agriculture, the risks inherent in an oligopolistic food system, the ways people of color and Indigenous people have been exploited by the food system, and the continuing toll that unhealthy food choices are taking on our health.

Most hopefully, however, *Food, Inc. 2* also explores the emerging alternatives—technological, social, economic, and political—that are helping millions to discover more sustainable and healthier ways to feed their families and communities. Readers of this book will learn how they can contribute to the creation of a more just food system that will serve the needs of all humankind as well as the future of our planet.

The book is divided into three parts. Part I, "Wake-Up Calls," brings back Michael Pollan and Eric Schlosser, the two journalists whose work helped inspire the first *Food, Inc.* movie—and whose 2020 essays on the way the COVID-19 pandemic exposed the ugly underbelly of the food system helped convince filmmakers Robby Kenner and Melissa Robledo that the time was ripe for a sequel.

Part II, the heart of the book, includes thirteen essays by an array of experts who delve deep into specific aspects of our food system, explaining the dysfunctions from which it suffers and the kinds of reforms we need to pursue. Among others, you'll hear from the brilliant Brazilian scientist Carlos A. Monteiro about the dangers of what he calls "ultra-processed" foods (and the steps his country is taking to minimize those dangers); from dairy farmer Sarah E. Lloyd about the movement to liberate small farmers from the overweening political and economic power of "Big Ag"; from Black farmer Leah Penniman about the ongoing quest for food and land justice for marginalized peoples; from labor rights activist Saru Jayaraman and the Coalition of Immokalee Workers about the battle to secure fair treatment for those who bring us the food we eat; from US senator Cory Booker about what Washington needs to do to create sustainable food policies for our nation; and from David E. Kelley, Andrew Zimmern, Christiana Musk, and Michiel Bakker about the technological and social breakthroughs that are opening promising new pathways to healthier eating for all.

Some of the topics you'll read about in these chapters may be familiar, but I predict you'll find yourself making new discoveries about issues you thought you understood. And I think even the best-informed foodie will encounter new aspects of the food world that are eye-opening, intriguing, often disturbing, and sometimes inspiring.

Finally, in Part III, you'll learn about some of the things you can do to support the growing efforts to transform our food system. Danielle Nierenberg explores a variety of steps we can all take to nourish both people and the planet, and those seeking further information and ways to become engaged will also find a list, arranged by topic, of many of the leading organizations that are currently at the forefront of reform efforts.

In the years since the first *Food, Inc.* film and book appeared, I've been gratified to hear from so many viewers and readers who told me how deeply their lives had been impacted by the stories, insights, and ideas they'd received from those sources. I hope that you, too, will be excited and energized by the powerful messages our contributors have provided in the following pages.

Few things impact us as deeply as the food we eat every day. The experts in this book can help us find a better way to eat—for our sake, and for the sake of millions of our fellow humans around the world today and in generations to come.

Karl Weber
Irvington, New York

Part I | WAKE-UP CALLS

1.
The Sickness in Our Food Supply*

By Michael Pollan

For more than thirty years, Michael Pollan has been writing books and articles about the places where the human and natural worlds intersect: on our plates, in our farms and gardens, and in our minds. Pollan is the author of eight books, six of which have been *New York Times* bestsellers; three of them (including his latest, *How to Change Your Mind*) were immediate #1 *New York Times* bestsellers. Previous books include *Cooked, Food Rules, In Defense of Food: An Eater's Manifesto,* and *The Omnivore's Dilemma: A Natural History of Four Meals,* which was named one of the ten best books of 2006 by both the *New York Times* and the *Washington Post*.

This essay, originally published during the early months of the COVID-19 pandemic, was part of the inspiration behind the decision to launch production of the sequel to the 2008 film *Food, Inc.*

* The article was originally published in the *New York Review of Books*, May 12, 2020. Reprinted by permission of the author.

Only when the tide goes out," Warren Buffett observed, "do you discover who's been swimming naked." For our society, the COVID-19 pandemic represents an ebb tide of historic proportions, one that is laying bare vulnerabilities and inequities that in normal times have gone undiscovered. Nowhere is this more evident than in the American food system. A series of shocks has exposed weak links in our food chain that threaten to leave grocery shelves as patchy and unpredictable as those in the former Soviet bloc. The very system that made possible the bounty of the American supermarket—its vaunted efficiency and ability to "pile it high and sell it cheap"—suddenly seems questionable, if not misguided. But the problems the novel coronavirus has revealed are not limited to the way we produce and distribute food. They also show up on our plates, since the diet on offer at the end of the industrial food chain is linked to precisely the types of chronic disease that render us more vulnerable to COVID-19.

The juxtaposition of images in the news of farmers destroying crops and dumping milk with empty supermarket shelves or hungry Americans lining up for hours at food banks tells a story of economic efficiency gone mad. Today the United States actually has two separate food chains, each supplying roughly half of the market. The retail food chain links one set of farmers to grocery stores, and a second chain links a different set of farmers to institutional purchasers of food, such as restaurants, schools, and corporate offices. With the shutting down of much of the economy as Americans stay home, this second food chain has essentially collapsed. But because of the way the industry has developed over the past several decades, it's virtually impossible to reroute food normally sold in bulk to institutions to the retail outlets now clamoring for it. There's still plenty of food coming from American farms, but no easy way to get it where it's needed.

How did we end up here? The story begins early in the Ronald Reagan administration, when the Justice Department rewrote the rules of antitrust enforcement: if a proposed merger promised to lead to greater marketplace "efficiency"—the watchword—and wouldn't harm the consumer, i.e., didn't raise prices, it would be approved. (It's worth noting that the word "consumer" appears nowhere in the Sherman Antitrust Act, passed in 1890. The law sought to protect producers—including farmers—and our politics from undue concentrations of corporate power.)[1] The new policy, which subsequent administrations have left in place, propelled a wave of mergers and acquisitions in the food industry. As the industry has grown steadily more concentrated since the 1980s, it has also grown much more specialized, with a tiny number of large corporations dominating each link in the supply chain. One chicken farmer interviewed recently in *Washington Monthly*—who sells millions of eggs into the liquified egg market, destined for omelets in school cafeterias—lacks the grading equipment and packaging (not to mention the contacts or contracts) to sell his eggs in the retail marketplace.[2] That chicken farmer had no choice but to euthanize thousands of hens at a time when eggs are in short supply in many supermarkets.

On April 26, John Tyson, the chairman of Tyson Foods, the second-largest meatpacker in America, took out ads in the *New York Times* and other newspapers to declare that the food chain was "breaking," raising the specter of imminent meat shortages as outbreaks of COVID-19 hit the industry.[3] Slaughterhouses have become hot zones for contagion, with thousands of workers now out sick and dozens of them dying.[4] This should come as no surprise: social distancing is virtually impossible in a modern meat plant, making it an ideal environment for a virus to spread. In recent years, meatpackers have successfully lobbied regulators to increase line speeds, with the result that workers must stand shoulder to shoulder cutting and deboning animals so quickly that they can't pause long enough to cover a cough, much less go to the bathroom, without carcasses passing them by. Some chicken-plant workers, given no regular bathroom breaks, now wear diapers.[5] A worker can ask for a break, but the plants are so loud that he or she can't be heard without speaking directly into the ear of a supervisor. Until recently, slaughterhouse workers had little or no access to personal protective equipment; many of them were also encouraged to keep working even after exposure to the virus. Add to this the fact that many meat-plant workers are immigrants who live in crowded conditions with little or no health care, and you have a population at dangerously high risk of infection.

When the number of COVID-19 cases in America's slaughterhouses exploded in late April—12,608 confirmed, with forty-nine deaths as of

May 11—public health officials and governors began ordering plants to close. It was this threat to the industry's profitability that led to Tyson's declaration, which President Donald Trump would have been right to see as a shakedown: the president's political difficulties could only be compounded by a shortage of meat. In order to reopen their production lines, Tyson and his fellow packers wanted the federal government to step in and preempt local public health authorities; they also needed liability protection, in case workers or their unions sued them for failing to observe health and safety regulations.

Within days of Tyson's ad, President Trump obliged the meatpackers by invoking the Defense Production Act. After having declined to use it to boost the production of badly needed coronavirus test kits, he now declared meat a "scarce and critical material essential to the national defense." The executive order took the decision to reopen or close meat plants out of local hands, forced employees back to work without any mandatory safety precautions, and offered their employers some protection from liability for their negligence. On May 8, Tyson reopened a meatpacking plant in Waterloo, Iowa, where more than a thousand workers had tested positive.

The president and America's meat eaters, not to mention its meat-plant workers, would never have found themselves in this predicament if not for the concentration of the meat industry, which has given us a supply chain so brittle that the closure of a single plant can cause havoc at every step, from farm to supermarket. Four companies now process more than 80 percent of beef cattle in America; another four companies process 57 percent of the hogs. A single Smithfield processing plant in Sioux Falls, South Dakota, processes 5 percent of the pork Americans eat. When an outbreak of COVID-19 forced the state's governor to shut that plant down in April, the farmers who raise pigs committed to it were stranded.

Once pigs reach slaughter weight, there's not much else you can do with them. You can't afford to keep feeding them; even if you could, the production lines are designed to accommodate pigs up to a certain size and weight, and no larger. Meanwhile, you've got baby pigs entering the process, steadily getting fatter. Much the same is true for the hybrid industrial chickens, which, if allowed to live beyond their allotted six or seven weeks, are susceptible to broken bones and heart problems and quickly become too large to hang on the disassembly line. This is why the meat-plant closures forced American farmers to euthanize millions of animals at a time when food banks were overwhelmed by demand.[6]

Under normal circumstances, the modern hog or chicken is a marvel of brutal efficiency, bred to produce protein at warp speed when given the right food and pharmaceuticals. So are the factories in which they are killed and cut into parts. These innovations have made meat, which for most of human history has been a luxury, a cheap commodity available to just about all Americans; we now eat, on average, more than nine ounces of meat per person per day, many of us at every meal.[7] COVID-19 has brutally exposed the risks that accompany such a system. There will always be a trade-off between efficiency and resilience (not to mention ethics); the food industry opted for the former, and we are now paying the price.

Imagine how different the story would be if there were still tens of thousands of chicken and pig farmers bringing their animals to hundreds of regional slaughterhouses. An outbreak at any one of them would barely disturb the system; it certainly wouldn't be front-page news. Meat would probably be more expensive, but the redundancy would render the system more resilient, making breakdowns in the national supply chain unlikely. Successive administrations allowed the industry to consolidate because the efficiencies promised to make meat cheaper for the consumer, which it did. It also gave us an industry so powerful it can enlist the president of the United States in its efforts to bring local health authorities to heel and force reluctant and frightened workers back onto the line.

Another vulnerability that the novel coronavirus has exposed is the paradoxical notion of "essential" workers who are grossly underpaid and whose lives are treated as disposable. It is the men and women who debone chicken carcasses flying down a line at 175 birds a minute, or pick salad greens under the desert sun, or drive refrigerated produce trucks across the country who are keeping us fed and keeping the wheels of our society from flying off. Our utter dependence on them has never been more clear. This should give food and agricultural workers a rare degree of political leverage at the very moment they are being disproportionately infected.

Scattered job actions and wildcat strikes are beginning to pop up around the country—at Amazon, Instacart, Whole Foods, Walmart, and some meat plants—as these workers begin to flex their muscle.[8] This is probably just the beginning. Perhaps their new leverage will allow them to win the kinds of wages, protections, and benefits that would more accurately reflect their importance to society.

So far, the produce sections of our supermarkets remain comparatively well stocked, but what happens this summer and next fall, if the outbreaks that have crippled the meat industry hit the farm fields? Farmworkers, too, live and work in close proximity, many of them undocumented immigrants crammed into temporary quarters on farms. Lacking benefits like sick pay, not to mention health insurance, they often have no choice but to work even when infected. Many growers depend on guest workers from Mexico to pick their crops. What happens if the pandemic—or the Trump administration, which is using the pandemic to justify even more restrictions on immigration—prevents them from coming north this year?

The food chain is buckling. But it's worth pointing out that there are parts of it that are adapting and doing relatively well. Local food systems have proved surprisingly resilient. Small, diversified farmers who supply restaurants have had an easier time finding new markets; the popularity of community-supported agriculture (CSA) is taking off, as people who are cooking at home sign up for weekly boxes of produce from regional growers. (The renaissance of home cooking and baking is one of the happier consequences of the lockdown, good news both for our health and for farmers who grow actual food, as opposed to commodities like corn and soy.) In many places, farmers' markets have quickly adjusted to pandemic conditions, instituting social-distancing rules and touchless payment systems. The advantages of local food systems have never been more obvious, and their rapid growth during the past two decades has at least partly insulated many communities from the shocks to the broader food economy.

The pandemic is, willy-nilly, making the case for deindustrializing and decentralizing the American food system, breaking up the meat oligopoly, ensuring that food workers have sick pay and access to health care, and pursuing policies that would sacrifice some degree of efficiency in favor of much greater resilience. Somewhat less obviously, the pandemic is making the case not only for a different food system but for a radically different diet as well.

It's long been understood that an industrial food system built upon a foundation of commodity crops like corn and soybeans leads to a diet dominated by meat and highly processed food. Most of what we grow in this country is not food, exactly, but rather feed for animals and the building blocks from which fast food, snacks, soda, and all the other wonders of food processing, such as high-fructose corn syrup, are manufactured. While some sectors of agriculture are struggling during the pandemic, we can expect the corn and soybean crop to escape more or less unscathed. That's because it takes remarkably little labor—typically a single farmer on a tractor, working alone—to plant and harvest thousands of acres of these crops. So processed foods should be the last kind to disappear from supermarket shelves.

Unfortunately, a diet dominated by such foods (as well as lots of meat and little in the way of vegetables or fruit—the so-called Western diet) predisposes us to obesity and chronic diseases such as hypertension and type 2 diabetes. These "underlying conditions" happen to be among the strongest predictors that an individual infected with COVID-19 will end up in the hospital with a severe case of the disease. The Centers for Disease Control and Prevention have reported that 49 percent of the people hospitalized for COVID-19 had preexisting hypertension, 48 percent were obese, and 28 percent had diabetes.[9]

Why these particular conditions should worsen COVID-19 infections might be explained by the fact that all three are symptoms of chronic inflammation, which is a disorder of the body's immune system. (The Western diet is by itself inflammatory.) One way that COVID-19 kills is by sending the victim's immune system into hyperdrive, igniting a "cytokine storm" that eventually destroys the lungs and other organs. A new Chinese study conducted in hospitals in Wuhan found that elevated levels of C-reactive protein, a standard marker of inflammation that has been linked to poor diet, "correlated with disease severity and tended to be a good predictor of adverse outcomes."[10]

A momentous question awaits us on the far side of the current crisis: Are we willing to address the many vulnerabilities that the novel coronavirus has so dramatically exposed? It's not hard to imagine a coherent and powerful new politics organized around precisely that principle. It would address the mistreatment of essential workers and gaping holes in the social safety net—including access to health care and sick leave—which we now understand, if we didn't before, would be a benefit to all of us. It would treat public health as a matter of national security, giving it the kind of resources that threats to national security warrant.

But to be comprehensive, this post-pandemic politics would also need to confront the glaring deficiencies of a food system that has grown so concentrated that it is exquisitely vulnerable to the risks and disruptions now facing us. In addition to protecting the men and women we depend on to feed us, it would seek to reorganize our agricultural policies to promote health rather than mere production, by paying attention to the quality as well as the quantity of the calories it produces. For even when our food system is functioning "normally," reliably supplying the supermarket shelves and drive-throughs with cheap and abundant calories, it is killing us—slowly in normal times, swiftly in times like these. The food system we have is not the result of the free market. (There hasn't been a free market in food since at least the Great Depression.) No, our food system is the product of agricultural and antitrust policies—political choices—that, as has suddenly become plain, stand in urgent need of reform.

2.

The Essentials: How We're Killing the People Who Feed Us*

By Eric Schlosser

Eric Schlosser is the author of the *New York Times* bestsellers *Fast Food Nation, Reefer Madness*, and *Command and Control*, a finalist for the Pulitzer Prize in History. He is a contributing writer at the *Atlantic*, where his work has been published for almost thirty years. Schlosser was an executive producer of the feature films *Fast Food Nation* and *There Will Be Blood*. He has also helped to produce documentaries about the food system, the plight of American ranchers, the abuse of migrant farmworkers, and the threat posed by nuclear weapons: *Food, Inc., Hanna Ranch, Food Chains*, and *Command and Control*. He cocreated *the bomb*, a multimedia installation performed at the Tribeca film festival, the Berlin film festival, and the Nobel Peace Prize ceremonies. Two of his plays have been staged in London: *Americans* at the Arcola Theatre and *We the People* at Shakespeare's Globe. His next book is about American prisons.

* Originally published in the *Atlantic*, May 12, 2020. Reprinted by permission of the author. The update at the conclusion of the essay was written in April 2023 especially for this book.

Under the headline "A Delicate Balance: Feeding the Nation and Keeping Our Employees Healthy," a letter from John Tyson appeared as a full-page ad in the *New York Times*, the *Washington Post*, and the *Arkansas Democrat-Gazette* on Sunday, April 26, 2020. Tyson is chairman of the board at Tyson Foods, the largest American-owned meatpacking company, and the grandson of its founder. "In small communities around the country, where we employ over 100,000 hard-working men and women, we're being forced to shutter our doors," he wrote. "This means one thing—the food supply chain is vulnerable." He raised the prospect that millions of animals would have to be "depopulated" and that only a "limited supply of our products" would be available if Tyson had to close its slaughterhouses.

What John Tyson failed to mention is that meatpacking plants, along with prisons, had become the nation's leading hot spots for the spread of COVID-19 infections. Thousands of meatpacking workers had fallen ill, many had died, and local health departments were considering whether to shut down plants operated by the industry giants: Tyson, Cargill, Smithfield Foods, and JBS USA. Two days after the publication of Tyson's letter, President Donald Trump issued an executive order that declared meatpacking plants to be "critical infrastructure" under the Defense Production Act of 1950 and prohibited their closure by state health authorities. The order provided meatpacking companies with a legal defense from liability claims by their employees. But it failed to impose any federal rules on how those companies must protect workers from outbreaks of COVID-19 at meatpacking plants.

By issuing that order, Trump helped an industry that has long been a strong supporter of the Republican Party. He reduced the likelihood that meat prices would greatly increase in the months leading up to the 2020 presidential election. And he confirmed what critics of the large meatpackers have said for years: some of these companies care more about profits than the lives of their workers, the well-being of the communities where they operate, and the health of the American people. Adding insult to injury, Kim Reynolds, the governor of Iowa—where major outbreaks of COVID-19 have been linked to meatpacking plants—announced that slaughterhouse employees who refuse to show up for work will be ineligible for unemployment benefits. While running for governor, Reynolds accepted hundreds of thousands of dollars in campaign funding from donors close to the meat industry.

One line in John Tyson's letter stood out to me: "Tyson Foods places team member safety as our top priority." A few weeks earlier, Andre Nogueira, the president of JBS, commenting on an outbreak of COVID-19 at a JBS slaughterhouse in Greeley, Colorado, said that his company's first priority "is our team members' safety." Not long afterward, Keira Lombardo, executive vice president of corporate affairs and compliance at Smithfield, said, "The health and safety of our employees is our top priority at all times." Following the release of Trump's executive order, Julie Anna Potts, president of the North American Meat Institute (NAMI), joined the chorus: "The safety of the heroic men and women working in the meat and poultry industry is the first priority."

Anyone who's spent time in an American meatpacking community would find those assertions laughable if the truth weren't so tragic and heartbreaking.

In recent weeks, workers at Tyson, JBS, and Smithfield haven't felt like their safety was a top priority. Before the widespread publicity about outbreaks at meatpacking plants, they'd routinely been denied face masks, social distancing, paid sick leave, and information about the number of COVID-19 infections in their workplace.

"Until someone dies, all these mediocre measures are being taken," Billy Williams, a union steward at a Tyson pork plant in Logansport, Indiana, told the *Pharos-Tribune*. "I'd rather have somebody go without their bacon and have my coworkers alive."

Crystal Rodriguez, a single mother with four children who works at the JBS plant in Greeley, Colorado, expressed frustration in an interview with CPR News. "I'm kind of angry because I don't understand why everybody's lives are being put at risk just to make the product," she said about policies at the slaughterhouse, where six workers have died and more than two hundred have been sickened with COVID-19. "There's never any soap in the bathroom so we can wash our hands." She later tested positive for the virus—and Sergio Rodriguez, her fifty-eight-year-old father, also an employee at the JBS plant, has spent nearly a month in the intensive care unit at a local hospital, fighting COVID-19.

When asked about Rodriguez's accusations, Nikki Richardson, a spokesperson for JBS, responded that the absence of soap, if true, would have been a violation of company policy, which requires that "all sanitizing and cleaning supplies are readily available at all times." Richardson added that employees must now wear surgical masks on company property at all times, that JBS does not want sick employees coming to work, that it will not punish absences for health reasons, and that workers afraid for their health can call the company and receive unpaid sick leave. "Throughout this pandemic, the health and safety of our team members has been and remains our highest priority," Richardson wrote to me in an email.

Under the pseudonym Jane Doe, a worker at a Smithfield pork plant in Milan, Missouri, joined a lawsuit against the company on April 23. She wasn't seeking any money—she wanted a court order that would force Smithfield to obey coronavirus guidance from public health officials and the Centers for Disease Control and Prevention (CDC). "I am afraid for my health and safety," she said in the lawsuit, "as well as the health and safety of people I am in contact with, and the larger community because of the way in which Smithfield is managing the plant in response to COVID-19." On May 5, US district judge Greg Kays dismissed the lawsuit. Appointed to the bench by President George W. Bush, Kays wrote that Smithfield had taken significant steps to protect workers and that two federal agencies, the US Department of Agriculture (USDA) and the Occupational Safety and Health Administration (OSHA), were responsible for ensuring compliance with guidelines. Kays also absolved the company: "No one can guarantee health for essential workers—or even the general public—in the middle of this global pandemic."

Smithfield said after the ruling: "From the start, we stated that this lawsuit was frivolous, full of specious allegations that were without factual or legal merit, and that the assertions were based on speculation, hearsay, anonymous declarations, and outdated information. This was nothing more than an attempt by a number of interconnected groups to promote their agenda through outrageous accusations."

When I inquired about Smithfield's overall response to the coronavirus, a publicist replied by sending links to the company's website.

Cargill did not respond to a request for comment.

Gary Mickelson, the director of media relations at Tyson, told me in an email that workers who test positive for COVID-19 now remain on sick leave until they are no longer contagious and will receive 90 percent of their normal pay until June 30. Moreover, Tyson has "put in place enhanced safety precautions and installed protective social distancing measures throughout all facilities." He also disputed the notion that Tyson's plants have become hot spots for COVID-19, suggesting that the high rate of testing at its facilities offers a misleading basis of comparison. Rates of infection may be comparable or higher elsewhere, though undetected. "The health and safety of our team members, their families, and communities is our top priority," Mickelson wrote.

Mickelson's suggestion that higher rates of testing at Tyson plants have created a misperception about them—and are not indicative of higher rates of COVID-19—does not correspond to the facts. At the Tyson pork plant in Waterloo, Iowa, more than one-third of the workers have tested positive for COVID-19. Based on recent antibody testing, that rate is about 50 percent higher than the proportion of people in New York City, the epicenter of the national outbreak, who have been sickened by the disease.

In conservative circles, a different argument has emerged: meatpacking workers are responsible for their own illnesses. "Living circumstances in certain cultures are different than they are with your traditional American family," a Smithfield spokesperson told *BuzzFeed News*—a comment that the company later disavowed. Wisconsin chief justice Patience Roggensack dismissed the spread of COVID-19 in Brown County, Wisconsin, home to a JBS plant, saying the workers who'd fallen ill weren't "regular folks." According to *Politico*, Alex Azar, the secretary of health and human services, told a group of lawmakers that workers were unlikely to be infected at meatpacking plants and that their "home and social" habits were spreading the virus. South Dakota governor Kristi Noem may have been the first Republican to express that view publicly. "We believe that 99 percent of what's going on today wasn't happening inside the facility," Noem told Fox News on April 13, while discussing an outbreak at a Smithfield pork plant where hundreds of workers had tested positive. "It was more at home, where these employees were going home and spreading some of the virus, because a lot of these folks that work at this plant live in the same community, the same building, sometimes the same apartments."

However compelling that argument may seem to the industry, it does not explain why three USDA food-safety inspectors who oversee meat-packing plants have died from COVID-19 and almost two hundred have been sickened. Of course, it's possible that their home and social habits were not "regular." A more likely explanation is that, in the early days of the pandemic, the USDA's Food Safety and Inspection Service not only failed to give protective equipment to its inspectors but also prohibited them from wearing masks inside meatpacking plants—concerned that the wrong message might be sent about the risk of COVID-19. On April 9, the agency said that inspectors could wear masks on the job, if the meatpacking company that owned the plant gave them permission to do so. Inspectors were encouraged to find their own masks and promised a $50 reimbursement for "the purchase of face coverings or materials to make face coverings." One month later, after meatpacking plants had been widely criticized as hot spots for spreading COVID-19, the USDA finally began to provide masks to its inspectors. "The safety and well-being of our employees is our top priority," a USDA spokesman said.

More than twenty years ago, a former meatpacking worker in Amarillo, Texas, told me the priority more important than anything else at an American slaughterhouse, the priority that still comes first today: "The chain will not stop." When a plant is up and running, fully staffed, the more meat it can process that day, the more profitable it will be. The speed of production and the amount of revenue are inextricably linked. Whenever possible, worker injuries aren't allowed to slow the throughput. "I've seen bleeders, and they're gushing because they got hit [by a knife] right in the vein, and I mean, they're almost passing out," she said, "and here comes the supply guy again, with the bleach, to clean the blood off the floor, but the chain never stops. It never stops."

The industry practice of making hundreds of workers stand close together at a production line—with sharp knives and a fast line speed—endangers not only their safety but also food safety and public health. If mistakes are made, workers can get hurt and meat can get contaminated.

The huge processing facilities run by America's meatpacking companies are excellent vectors for spreading lethal strains of *E. coli*, antibiotic-resistant *Salmonella*, antibiotic-resistant *Staphylococcus aureus*, and now COVID-19.

Nevertheless, much to the industry's delight, the Trump administration has let the number of inspectors at OSHA fall to the lowest level in almost half a century. According to the National Employment Law Project, more than 40 percent of the top leadership positions at OSHA are currently unfilled. Last September, the USDA reduced food-safety inspections at some pork plants and gave them permission to increase maximum line speeds. The National Pork Producers Council hailed the changes and claimed that their impact would be "enhancing safety, quality, and consistency." While the coronavirus spread through American slaughterhouses last month, the USDA introduced a similar program that will enable Tyson and other poultry companies to speed up their production lines as well.

Meatpacking companies don't want their workers to be injured or sickened on the job. But they also don't want to spend the money necessary to reduce the extraordinary rate of those injuries and illnesses. And they especially don't want to pay the health-care costs of injured workers. A 2015 investigation by ProPublica found that, for the past few decades, Tyson has led a nationwide effort to make it harder for workers hurt on the job to receive benefits from workers' compensation plans: "Tyson self-insures, meaning it pays nearly all of its claims from its own pocket. When workers are injured, they're usually sent to a Tyson nurse at the plant. Their claims are processed by Tyson adjusters. And in many states, the company even has its own managed-care unit, handpicking the doctors that workers can see and advising those doctors on light-duty jobs injured employees might be able to do." In Texas, where private employers are not required to carry workers' compensation insurance, Tyson has opted out of the state system completely.

When a worker gets injured at the Tyson beef slaughterhouse in Amarillo, Texas, in order to get medical care from the company, that person must first sign a document saying: "I hereby voluntarily release, waive, and forever give up all my rights, claims, and causes of action, whether now existing or arising in the future, that I may have against the company, Tyson Foods, Inc., and their parent, subsidiary and affiliated companies and all of their officers, directors, owners, employees, and agents that arise out of or are in any way related to injuries (including a subsequent or resulting death) sustained in the course of my employment with the company."

If the injured worker doesn't sign the waiver, that person can be fired—and then has to file a lawsuit against Tyson to get any payments for medical bills. It's a fight that an immigrant worker is unlikely to win against a multinational corporation with annual revenues of about $40 billion. The Texas legislature passed a law in 2005 giving injured workers ten days to decide whether to sign such a waiver and hand over total control of their health care to their employer. Before that law was passed, meatpacking workers were sometimes asked to sign a waiver immediately after an injury. The pressure to sign was enormous. When a worker named Duane Mullin had both of his hands crushed in a hammer mill at the Amarillo slaughterhouse now owned by Tyson, a manager employed by its previous owner persuaded him to sign the waiver with a pen held in his teeth.

"Tyson Foods places team member safety as our top priority" belongs in the same category of plausibility as Trump's remark that "we did all the right moves" in handling the coronavirus pandemic—and Jared Kushner's description of the federal government's response to COVID-19 as a "great success story." A few facts offer a useful perspective. South Korea detected its first case of COVID-19 on January 20, and the United States detected its first case the following day. According to the latest figures, 695 people have tested positive for COVID-19 in Seoul, which has a population of 9.8 million. And 890 workers have tested positive at the Tyson pork plant in Logansport—more than one-third of the 2,200 workers at the plant.

Some grocery stores are now limiting how much meat customers can purchase, and Wendy's has been running out of ground beef. These shortages hardly qualify as a national emergency. During the Second World War, government rationing limited weekly meat purchases to about two pounds a person. Today the typical American consumes about twice that amount every week. If the Greatest Generation could defeat Nazi Germany and the empire of Japan on a smaller ration of meat, we can certainly eat less of it for the time being to spare the lives of meatpacking workers and their communities.

Cattle ranchers, hog farmers, and poultry growers deserve compensation for the livestock being euthanized because of the slowdowns at slaughterhouses. But the coronavirus isn't responsible for the problem. Hogs can live six to eight years in the wild and twice as long when they're domesticated. The fact that hundreds of thousands may have to be culled and discarded is one more sign that our centralized, industrialized food system isn't sustainable, lacks resilience, defies logic, and must be transformed.

Gettysburg was the deadliest battle of the Civil War. Over the course of three days in July 1863, almost eight thousand Confederate and Union soldiers were killed. The number of American deaths attributed to COVID-19 was about 8,500 over the course of two days this April. President Abraham Lincoln wrote an incomparably beautiful and powerful speech to honor the fallen at Gettysburg, urging that "we here highly resolve that these dead shall not have died in vain—that this nation, under God, shall have a new birth of freedom." A similar commitment should be made on behalf of the roughly eighty thousand Americans who have died from COVID-19 since late January. Those Americans have been disproportionately elderly, poor, and people of color, too often working at low-paying jobs deemed necessary for the rest of us: meatpacking worker, restaurant worker, farmworker, delivery person, grocery clerk.

Here are the essential things that we must achieve, the very least we must do to give those deaths meaning and take care of the people who feed us:

- A minimum wage that's a living wage for all workers—at least $15 an hour—and elimination of the subminimum wage for restaurant workers, which can be as low as $2.13 an hour. Health insurance for every single American. A safe workplace and fair compensation for every worker injured, sickened, or sexually harassed on the job.
- Accountability for workplace injuries. An accident is when you walk down the street, step on a banana peel, slip, and hurt your back. When thousands of meatpacking workers are suffering the

same kinds of amputations, lacerations, and cumulative-trauma injuries every year, those aren't industrial accidents. They're a business decision. Large fines should be imposed for workplace injuries, and criminal charges should be filed against the executives who consistently ignore them.

- Protection of the right to organize labor unions in every state and in every workplace, including at the franchised restaurants controlled by McDonald's and the other fast-food chains.
- Food free from contamination and adulteration, guaranteed by a food-safety system that hasn't been privatized. The federal government should severely punish companies that knowingly spread dangerous, antibiotic-resistant pathogens.
- Strict antitrust enforcement that will rid the food system of monopoly and monopsony power, ensure competition, and encourage the innovation that free-market forces produce. And amnesty for the millions of undocumented immigrants in the United States, the backbone of our food system, who must be given a pathway toward legal status.

Those are the essentials, and many more necessary reforms can be added to them.

"We are living in a failed state," George Packer eloquently argued in the *Atlantic* recently, outlining the many ways our political culture and governmental institutions have been corrupted. But in at least one respect, the exercise of power is now remarkably efficient and effective. As the efforts of the meatpacking industry demonstrate, to paraphrase Lincoln, today we have a government of big corporations, by big corporations, for big corporations.

And if we don't take action, and protest, and organize, and make sure to vote this November, that's what it will remain.

UPDATE

On May 12, 2022, the House Select Subcommittee on the Coronavirus Crisis issued a report entitled *"Now to Get Rid of Those Pesky Health Departments!": How the Trump Administration Helped the Meatpacking Industry*

Block Pandemic Worker Protections. The subcommittee had obtained more than 150,000 documents from the largest meatpacking companies, the federal agencies designated to regulate them, and the trade organizations fighting those regulations. The quote in the title came from an email written by Ashley Peterson, senior vice president of scientific and regulatory affairs at the National Chicken Council (NCC). During the spring of 2020, when COVID-19 was at its deadliest and effective countermeasures were being promoted—like routine testing, social distancing, and the use of face masks—Peterson agreed with a meatpacking executive that taking the temperature of workers was "all we should be doing" and that public health departments were the problem.

The report's conclusions were damning and blunt: "Internal meatpacking industry documents reviewed by the Select Subcommittee now illustrate that despite awareness of the high risks of coronavirus spread in their plants, meatpacking companies engaged in a concerted effort with Trump administration political officials to insulate themselves from coronavirus-related oversight, to force workers to continue working in dangerous conditions, and to shield themselves from legal liability for any resulting worker illness or death."

The House subcommittee found that the executive order issued by President Trump on April 28, 2020, had been conceived two weeks earlier by Ken Sullivan, the chief executive officer of Smithfield Foods, and Noel White, the chief executive officer of Tyson Foods. After the head of the nation's largest pork company consulted with the head of its largest chicken company, the idea was shared with top executives at other meatpacking companies, along with a warning about "positive cases, fear-driven absenteeism, and disincentives to work." The first draft of Trump's executive order was written by the legal department at Tyson Foods. Meatpacking executives had numerous conversations about the need for an executive order with staff members at the White House. Sullivan and White discussed it on the phone with Mark Meadows, the White House chief of staff. Julie Anna Potts, president of the North American Meat Institute, explained the purpose of the executive order in a private email to industry representatives: "(1) directing state and local authorities to keep plants open and (2) protection from liability for worker illnesses." As noted in my *Atlantic* article, Potts described the aim of the executive order somewhat differently in public: "The safety of the heroic men and women working in the meat and poultry industry is the first priority."

The House subcommittee found that Ken Sullivan, the head of Smithfield, obtained an early draft of recommendations to halt the spread of COVID-19 that the CDC was planning to issue. Sullivan marked up the draft by hand with his criticisms and sent it to Gregory Ibach, an undersecretary of the USDA appointed by Trump. "We are on it," Ibach replied. "The changes that were ultimately made to the CDC recommendations came at the behest of Smithfield," the subcommittee report said, "and were made by the Trump-appointed CDC director [Robert Redfield] over the objection of career officials." Safety measures like social distancing became voluntary. Redfield later claimed that the changes were justified by the "substantial protein shortage" that America would confront if production was slowed at its meatpacking plants.

The House subcommittee found that the risk of severe meat shortages, cited repeatedly by the industry to avoid worker protections, was a myth. For example, during the spring of 2020, the United States had 622 million pounds of pork in cold storage—enough to supply American grocery stores for more than a year. "During the first three quarters of 2020, Smithfield exported 90 percent more pork to China than it did during the same period in 2017, while JBS appears to have exported a whopping 370 percent more," the subcommittee report said. The profits of the meatpacking industry skyrocketed during the pandemic. Between 2020 and 2021, Tyson's net income increased from $2 billion to $3 billion, while the net income of JBS rose from $937 million to $4.2 billion.

The House subcommittee found that the health of meatpacking workers, their families, and their communities was deemed less important than those profits. The absence of social distancing at meatpacking plants made them ideal vectors for the spread of the coronavirus. During the spring of 2020, the number of COVID-19 cases in rural counties with meatpacking plants was ten times the number in rural counties without those facilities. And eight of the ten rural counties with the highest rates of COVID-19 infection in the United States had meatpacking plants.

A lawsuit filed against Tyson Foods in May 2022 suggests that the company was well aware of the dangers that its American workforce faced during the most lethal phase of the pandemic—and that it knew how to reduce them. On January 11, 2020, China reported its first confirmed death from COVID-19. Within weeks, Tyson Foods had halted or slowed production at some of its plants in China to prevent the spread of the novel coronavirus. According to the lawsuit:

> By February, all of Tyson's China-based operations and facilities had implemented and were following effective COVID-19 protocols that included providing masks and other appropriate PPE [personal protective equipment] to employees, checking employees' temperature twice a day, thoroughly educating employees on how to protect themselves from the spread of the virus, installing air filtration systems, establishing quarantine observation areas for potentially infected employees, restricting access to facilities (including symptomatic employees), and preventing employees from gathering in cafeterias and break rooms.

In early February, the vice president of food safety and quality management of Tyson Foods in China shared the coronavirus protocols of its Chinese plants with Tyson executives in the United States. In March, Tyson Foods suspended business travel for its American executives and allowed them to work remotely from home. The following month, Tyson spearheaded the effort to get President Trump to issue an executive order that would impede health departments, keep its American workers on the job, and limit the liability for their illnesses.

Representatives of Tyson Foods, Smithfield Foods, the National Chicken Council, and the North American Meat Institute declined to speak with me about the House subcommittee report—or anything else.

A Tyson representative told me via email that the company has spent more than $800 million to prevent the spread of COVID-19 at its American facilities, provided face masks to all its workers, increased their pay and bonuses, installed physical barriers between work stations, and required that its entire workforce be fully vaccinated. "Our top priority has been and continues to be the health and safety our team members," the representative wrote.

The Smithfield representative declined to answer any questions.

The National Chicken Council representative sent me its response to the House subcommittee report. "We regret that this report failed to shine light on the momentous efforts between industry, government and state and local health officials to keep employees safe and to keep Americans fed during one of the most challenging and uncertain times in our nation's history," the press release said. "NCC stands by its actions and those of its members during the pandemic."

The North American Meat Institute representative sent me its response to the House subcommittee report. "The report ignores the rigorous and comprehensive measures companies enacted to protect employees," Julie Anna Potts, the president of NAMI, said. "As more became known about the spread of the virus, the meat industry spent billions of dollars to reverse the pandemic's trajectory, protecting meat and poultry workers while keeping food on Americans' tables and our farm economy working." I was also sent a NAMI white paper on the COVID-19 pandemic, which includes this assertion: "The meat and poultry industry's most valuable asset is its workforce."

Amid shortages of immigrant workers, the meatpacking industry has recently been caught hiring a type of employee forbidden by law: children. In February 2023, the US Department of Labor found that at least one hundred children, aged thirteen to seventeen, worked for a company called Packers Sanitation Services at more than a dozen meatpacking plants in eight states. Seven of the children worked at Tyson plants. That same month, Republican legislators in Iowa introduced a bill allowing children as young as fourteen to work legally in meatpacking plants. Although the bill would prohibit child labor on the slaughterhouse floor, it would permit children to work in coolers at meat plants, expand the number of hours that children can work, and limit the liability of companies whose child laborers are sickened, injured, or killed on the job. In March 2023, Arkansas relaxed some of its restrictions on child labor, and seven other states are now considering similar legislation.

I asked representatives from Tyson Foods and the North American Meat Institute whether state laws should be amended to allow child labor at meatpacking plants.

Neither answered the question.

Eighty-five years ago, the Fair Labor Standards Act of 1938 prohibited child labor, placed limits on overtime, and established a federal minimum wage. It's worth remembering how American workers were treated before Congress passed that law. They were often forced to work ten to twelve hours a day, six days a week. They earned as little as $2.50 an hour, in today's dollars. They were sometimes paid in scrip, redeemable only at a company store, instead of money. About one-quarter of American children worked for sixty or more hours a week. The median wage for children working those hours was about $80 a week, in today's dollars. The US Supreme Court repeatedly overturned minimum wage laws and child labor laws in the name of "freedom."

The Supreme Court finally changed course in 1937, upholding a minimum wage enacted in Washington State. "The community is not bound to provide what is in effect a subsidy for unconscionable employers," Chief Justice Charles E. Hughes declared. The door had been opened to federal protections for workers, and President Franklin Delano Roosevelt soon urged support for the Fair Labor Standards Act. "No business which depends on paying less than living wages to its workers has any right to continue in this country," Roosevelt said.

That simple truth has been forgotten. The federal minimum wage hasn't been increased since 2009, the longest period without a raise since 1938. And worker rights are now being eroded in ways not seen since the early twentieth century. The meatpacking industry's response to the coronavirus offers a textbook case of how government agencies can be corrupted, how the public interest can be sacrificed for profit, and how the law can be subverted by wealthy private interests. Most of all, it serves as a reminder of a hard lesson learned more than a century ago: monopoly power is the opposite of freedom.

Part II | MAPPING THE ISSUES— AND THE ANSWERS

3.
Food, Cooking, Meals, Good Health, and Well-Being

By Carlos A. Monteiro and Geoffrey Cannon

Carlos A. Monteiro, MD, PhD, is a professor of nutrition and public health at the School of Public Health, University of São Paulo, Brazil, where he created the Center for Epidemiological Studies in Health and Nutrition. His research lines include methods and tools in population dietary assessment, epidemiology of all forms of malnutrition, dietary determinants of chronic diseases, and food processing and human health. On these subjects, he has published more than 250 journal articles that have more than twenty thousand citations in the Web of Science. He led the team responsible for the technical content of the 2014 Brazilian dietary guidelines. He has served on numerous national and international nutrition expert panels and committees and, since 2010, he has been a member of the World Health Organization (WHO) Nutrition Guidance Expert Advisory Group. In 2010, he received the Pan American Health Organization Abraham Horwitz Award for Excellence in Leadership in Inter-American Health.

Geoffrey Cannon works with the Center for Epidemiological Studies in Health and Nutrition, University of São Paulo, Brazil.

He was on the team responsible for the technical content of the 2014 Brazilian official national nutrition guidelines. From 2010 to 2016, he was editor of the journal *World Nutrition*. He is in the Web of Science top 1 percent of scholars of all subjects cited in high-impact journals. From 2000 to 2002, he worked at the Brazilian federal department of health. He drafted the first official Brazilian nutrition guidelines published in 2006 and was a member of the Brazilian government delegation to the WHO Executive Board meeting in January 2001. In Britain until 2000, he was director of the World Cancer Research Fund 1997 report *Food, Nutrition and the Prevention of Cancer* and chief editor of its second 2007 report. Books include *Superbug, The Politics of Food,* and *Dieting Makes You Fat.*

The most cogent idea of a healthy diet has changed. Now it is no longer, as implied in most current official dietary guides, just a matter of personally choosing and consuming foods containing amounts and types of nutrients that promote growth and prevent specified diseases. Instead, a much broader view of the relationship between food and health is required.

Here are six reasons why the current narrow view is inadequate.

1. Good health and its expression as well-being is more than physical growth and absence of illness. Good health and well-being is mental, emotional, and, many would say, spiritual, and it promotes useful, rewarding, and purposeful lives. Therefore, good food should be nourishing in all senses.

2. Focusing on nutrients is not enough. Whole foods contain thousands of bioactive compounds, many now known or believed to protect health, that are not classified as nutrients. At the same time, an increasing number of manufactured food products contain xenobiotics—novel chemical substances that are foreign to human (and animal) bodies. Both of these are generally ignored in conventional dietary guides.

3. Humans are naturally social, not isolated individuals. We usually grow up with parents and in families and live with partners, friends, and colleagues in communities, societies, and countries with their own customs and cultures. These all shape and sustain dietary patterns. Commensality—eating together—is a natural part of family and social life.

4. What most people eat is largely determined by what is available, attractive, and affordable. Depending on resources, choices of what to buy and eat are more or less constrained.

5. Decent public policies care for future generations.

6. Humans are part of the living and natural world and the biosphere. This means that our relationship to the food we eat must involve caring for the future of our species as well as the future of the environment on which we depend. These are abused and damaged by exploitation that diminishes unrenewable resources, disrupts climate, and distributes pollution. Diets high in meat and fish are likely to be nutrient dense. But meat consumption is now too high in many countries, so that much beef production now involves vast feedlots that wreck farmland, emit noxious waste, and destroy forests and savanna. General increased fish consumption, as commonly advised, would accelerate the diminution and even threaten the extinction of ocean fish stocks. Intensive growing of various plant foods makes prodigal use of energy, land, and water without threatening species.

Though these points may seem obvious and are increasingly accepted in this age of the Anthropocene—in which human beings determine the fate of the planet—acting on them, especially in combination, may seem complex. Here is a master solution, for people as consumers, citizens, parents, members of families, partners, friends, and colleagues, and for health professionals, scholars, and policymakers at all levels: recognize, valorize, and emphasize long-established dietary patterns based on a variety of whole or minimally processed foods, rather than nutrients and specific foods.

Below is outlined what this means for societies, cultures, and resources and for public policy planning and action at all levels. And personally? Simple. Enjoy freshly prepared and cooked meals daily, when
possible in company.

WHAT MAKES A HEALTHY DIET?

Sustained good health is commonly agreed to be the most desirable aspect
of life, and the nature and quality of food is known to be a crucial deter-
minant of health. But what makes a healthy diet?

Official dietary guidelines and other forms of guidance published
all over the world—such as those issued every five years in the United
States (the Dietary Guidelines for Americans, or DGAs)—assume that
healthy diets promote growth and prevent various diseases by containing
adequate dietary energy and chemical constituents identified as essential
macro- and micronutrients but are not excessive in energy, total fat, saturated fat, sugar, or sodium. Such guides are "nutricentric," meaning that
they group foods according to their relative contents of macronutrients
(protein, fat, and carbohydrate) and micronutrients (vitamins and minerals). At the time of writing, the twenty members of the 2025–2030 DGA
advisory committee are in session, and these new DGAs are due to be
issued in 2024.[1]

When these guides were first promulgated by governments, they
looked to be rational and successful. The food industries responded by
"fortifying" various products with vitamins and minerals, and from the
1970s and 1980s by manufacturing low-fat and low-saturated-fat options.
Rates of undernutrition and micronutrient deficiencies were reduced, especially in higher-income countries, and foods high in animal protein such
as cow milk, dairy products, and meat accelerated growth in early life. Reduced amounts of fat and saturated fat in common foods—produced and
advertised as low-fat by manufacturers—have commonly been thought to
have lowered rates of death from coronary heart disease, which had suddenly become epidemic in high-income countries.

So until the later 1980s, official dietary guidelines were agreed to be
successful.[2] The nutricentric concept of healthy diets that they embody
was rarely questioned by orthodox Western scholars.

But there is now no credible evidence that these dietary guides and the actions based on them—designed to reduce production and consumption of fat, saturated fat, sugar, and sodium in many manufactured foods—are effective. They are not preventing the ever-growing prevalence of obesity and diabetes, worst in the United States and other high-income countries but also now pandemic, or of related disorders and diseases. Nutricentric dietary guides have become failures.

THE SIGNIFICANCE OF PROCESSING AND
THE NOVA CLASSIFICATION

So what could work well? In 2010, a new concept of healthy and unhealthy diets was proposed in a commentary in *Public Health Nutrition* (*PHN*) with the title "Nutrition and Health: The Issue Is Not Food, nor Nutrients, So Much as Processing."[3]

The commentary pointed out that "almost all food is processed in some way." Foods have been preserved for thousands of years. Various forms of food processing are necessary, harmless, or benign, such as preservation of fruits and nonalcoholic fermentation. Other types—such as the partial hydrogenation that generates toxic trans-fatty acids, and various coal-tar dyes (many now withdrawn)—are harmful. Many others alter or transform the appearance or the nature of food. No food processes are inert. All are significant.

Despite this, "the issue of food processing is largely ignored or minimized in education and information about food, nutrition and health, and also in public health policies." The commentary's thesis was that the narrow nutricentric concept of diet has become inadequate and that the fundamental issue is what is done to foods between their natural state and when they are consumed, in particular their industrial processing.[4]

This thinking originated in Brazil, a very large country where food supplies, dietary habits, and disease patterns have shifted rapidly, as they have in many other middle-income and now also low-income countries. The idea was prompted by analyses of national Brazilian household food expenditure surveys.[5] These showed extreme reduction in the purchases of cooking and table oils and of fats, sugar, and salt—yet the prevalence of overweight and obesity was rapidly rising. The analysis also showed that the amounts of fat, free sugars, and sodium purchased were all increasing, not for use in preparing meals but in ready-to-consume products such as soft drinks, biscuits, sweet and salty snacks, and reconstituted meat products. In the period analyzed, while purchases of culinary ingredients and minimally processed rice and beans were dropping, purchases of biscuits had become twice as high and of soft drinks four times as high. The cause of this increase in unhealthy diets was obviously processing.

The 2010 commentary proposed a new food classification system, later named Nova, whereby all foods (including drinks) are categorized according to the extent and purpose of their processing. Special attention was given to those foods often vaguely described with terms such as "highly" or "heavily" processed; "fast," "industrial," or "convenience"; or in the popular press as "junk" or "fake." Michael Pollan in his 2008 book *In Defense of Food* was more precise.[6] He identified "edible food-like substances." These were described in another article in the 2010 *PHN* as "extractions and extrusions of refined fats, oils, starches and sugars, often with sprinklings of unrefined or lightly refined foods, sophisticated and constituted into products usually with preservatives and often with cosmetic additives—and increasingly often, synthetic vitamins and minerals."[7] These the *PHN* commentary named "ultra-processed foods."

The first national use of Nova was as the framework of the 2014 official *Dietary Guidelines for the Brazilian Population*.[8] These were compiled with the support of the World Health Organization and Pan American Health Organization, national representative bodies, and health professionals from all twenty-seven Brazilian states. They were published by the federal Ministry of Health in an initial edition of sixty thousand copies distributed to health, food, and nutrition centers at all levels—state, municipal, and local. The Nova system is also set out in a 2019 report issued by the UN Food and Agriculture Organization.[9]

Using Nova, the Brazilian guidelines separate all foods and drinks into four groups:

Group 1. Unprocessed and Minimally Processed Foods

Unprocessed or natural foods are the edible parts or products of plants (fruits, leaves, stems, seeds, roots), animals (flesh, organs, eggs, milk), fungi and seaweeds (after separation from nature), and also water.

Minimally processed foods are foods altered by processes that usually preserve the natural food structure (matrix) and do not add fat, sugar, or salt. These include removal of inedible or unwanted parts, drying, crushing, grinding, powdering, fractioning, filtering, roasting, boiling, nonalcoholic fermentation, pasteurization, chilling, freezing, placing in containers, and vacuum packaging.

Within Nova, the distinction between unprocessed and minimally processed foods is not especially significant. These processes are designed to preserve natural foods, to make them suitable for storage, or to make them safe, edible, or more pleasant to consume.

Unprocessed and minimally processed foods vary in energy density and in their content and balance of fats, carbohydrates, proteins, and their fractions and of vitamins, minerals, and other bioactive compounds. No single type of food can provide human beings with all necessary energy and essential nutrients in adequate balance, except for breast milk in the first six months of life.

In general, animal foods are good sources of proteins and various vitamins and minerals but contain little or no dietary fiber. Quite often they are energy dense and high in unhealthy types of fat. Plant foods are usually low in energy density and good sources of dietary fiber. Many are high in various micronutrients and other bioactive compounds, and some are good sources of proteins.

This is why the human species has evolved as omnivorous. It explains why a great variety of long-established food systems have in common the combination of plant foods with complementary nutrient profiles, such as grains (cereals) with legumes (pulses), or roots with legumes, or grains with vegetables, usually also with modest amounts of animal foods.

In appropriate varieties, amounts, and combinations, all foods in this group are the basis for healthy diets.

Group 2. Processed Culinary Ingredients

This group includes oils, butter, lard, sugar, honey, and salt. They are substances extracted from Group 1 foods or else from nature by processes such as pressing, refining, grinding, milling, and drying. Some methods used to make processed culinary ingredients are originally ancient. Now they are usually industrial products, designed to make durable products suitable for use in home, restaurant, and cafeteria kitchens to prepare, season, and cook unprocessed or minimally processed foods and turn them into freshly prepared dishes and meals.

In isolation, processed culinary ingredients are unbalanced, having been depleted in some or most nutrients. Other than salt, they are also energy dense, at four hundred or nine hundred calories per one hundred grams. This is around three to six times more than cooked grains and around ten to twenty times more than cooked vegetables.

But they are rarely if ever consumed by themselves. They are used in combination with foods to make palatable, diverse, nourishing, and enjoyable meals and dishes such as stews, soups, broths, salads, breads, preserves, drinks, and desserts. Thus, oils are used in the cooking of grains, vegetables, legumes, and meat and are added to salads. Table sugar is used to prepare fruit- or milk-based desserts. It is misleading to assess their nutritional significance in isolation. They should always be assessed in combination with foods.

Many culinary ingredients are cheap and can be overused. When used carefully and in small amounts, they result in delicious dishes and meals that are nutritionally balanced, with energy densities much lower than those of most ready-to-consume food products.

Group 3. Processed Foods

Processed foods are natural foods altered by processes that usually preserve the food matrix but include the addition of fat, sugar, or salt to the original food. They include canned or bottled vegetables or legumes preserved in brine, whole fruits preserved in syrup, tinned fish preserved in oil, most freshly baked breads, cheeses that are not mass-produced, and some animal foods to which only salt is added, such as dried meat and fish, smoked meat and fish, and pastrami.

Processes also include various preservation or cooking methods and, with breads and cheeses, nonalcoholic fermentation. Processing here increases the durability of Group 1 foods or modifies or enhances their sensory qualities. Most processed foods have two or three ingredients and are recognizable as modified versions of Group 1 foods. They are generally produced to be consumed as part of meals or dishes and also may be consumed by themselves as snacks. Most are highly palatable.

As with culinary ingredients, some methods used to make processed food products are originally ancient and are still used domestically or artisanally. Now, almost all are manufactured industrially.

Processed foods usually retain the basic identity and most constituents of the original food. But when excessive oil, sugar, or salt are added, they become nutritionally unbalanced. Except for canned vegetables, their energy density ranges from moderate (around 150 to 250 calories per one hundred grams for most processed meats) to high (around 300 to 400 calories per one hundred grams for most cheeses).

Like processed culinary ingredients, they can be overused. When used sparingly—in the case of processed meats, only occasionally—they also result in delicious dishes and meals that are nutritionally balanced with energy densities lower than those of most ready-to-consume products.

Thus, in appropriate variety and amounts, these three Nova groups are beneficial and healthy. Now for the fourth group.

Group 4. Ultra-Processed Foods

Ultra-processed foods are not modified whole foods. They are increasingly sophisticated formulations of ingredients, mostly of exclusive industrial use, typically created by a series of industrial techniques and processes (hence "ultra-processed").

Some common ultra-processed products are carbonated sweet and "diet" soft drinks; sweet, fatty, or salty packaged snacks; candies (confectionery); mass-produced packaged breads and buns, cookies (biscuits), pastries, cakes, and cake mixes; margarine and other spreads; sweetened breakfast "cereals"; "milk," "fruit," yogurt, and "energy" drinks; preprepared meat, cheese, pasta, and pizza dishes; poultry and fish "nuggets" and "sticks"; sausages, burgers, hot dogs, and other reconstituted meat products; powdered and packaged "instant" soups, noodles, and desserts; baby formula; and many other types of products, typically ready to eat, drink, or heat.

Processes enabling the manufacture of ultra-processed foods involve several steps and different industries. They start with the fractioning of whole foods into substances including sugars, oils, fats, proteins, starches, and fiber. These are mostly obtained from a few high-yield plant foods—such as corn, wheat, soy, cane, or beet—grown on specialized large farms and from crushing or grinding animal carcasses and remnants, usually from intensively reared livestock.

These substances are then subjected to hydrolysis, hydrogenation, or other chemical transformations and then are assembled with little if any whole food added, using industrial techniques such as extrusion, molding, and pre-frying. Colors, flavors, emulsifiers, and other additives are used to make the final product palatable or hyper-palatable, and then completed with attractive, usually synthetic packaging.

The sugar, oils and fats, and salt used to make processed foods are often ingredients of ultra-processed foods, commonly in combination. Additives that prolong product duration, protect original properties, and prevent proliferation of microorganisms are used in both processed and ultra-processed foods, as well as in processed culinary ingredients and sometimes in minimally processed foods.

Ingredients characteristic of ultra-processed foods are either food substances of no or rare culinary use or else classes of additives whose function is to make the final product sellable and, often, hyper-palatable.

Food substances rarely or never found in home kitchens that are used to formulate ultra-processed foods include varieties of sugars (fructose, high-fructose corn syrup, "fruit juice concentrates," invert sugar and syrup, maltodextrin, dextrose, and lactose), modified oils (hydrogenated or interesterified), and various sources of protein (hydrolyzed proteins, soy protein isolate, gluten, casein, whey protein, and "mechanically separated meat").

Classes of additives used only in the manufacture of ultra-processed foods include flavors, flavor enhancers, colors, emulsifiers, emulsifying salts, sweeteners, thickeners, and agents for foaming, anti-foaming, bulking, carbonating, gelling, and glazing. All of them, most notably flavors and colors, either disguise unpleasant sensory properties created by ingredients, processes, or packaging used in manufacture of ultra-processed foods or give the final product intensely attractive sensory properties. The purpose of cosmetic additives—to disguise combinations of chemically transformed substances as real food—is overlooked in conventional dietary guidance.

Processes and ingredients used for the manufacture of ultra-processed foods are designed to create highly profitable products: low-cost ingredients, long shelf life, powerful branding. Their convenience (imperishable, ready to consume), hyper-palatability, and ownership by transnational corporations using pervasive advertising and promotion give ultra-processed foods enormous market advantages. They are therefore liable to displace all other Nova food groups and replace freshly made regular meals and dishes with snacking anytime, anywhere.

Not all ultra-processed foods are recent. Some of the first such products enabled by mass industrialization have been commonly consumed for generations, though mostly in much smaller amounts than now. They include soft drinks, packaged cookies, meat and yeast extracts, ice cream, chocolates, packaged candies, margarines, and infant formulas.

Some of what are now ultra-processed foods were originally manufactured only from Nova Group 1 foods and salt, sugar, or other substances from Group 2, and thus would be classed in Group 3 as processed foods. But as now formulated most of them are ultra-processed. Examples are commercial wrapped breads, packaged cakes and pies, and preprepared animal products such as hot dogs and burgers. Packaged, ready-to-heat products consumed at home or at fast-food outlets such as meat, cheese, pizza and pasta dishes, and french fries may look much the same as home-cooked food, but their different formulations and the ingredients used in their pre-preparation make them ultra-processed.

THE FOOD INDUSTRY AND TRANSNATIONAL CORPORATIONS

National and global diets also need to be seen in a new light because, in recent decades, the nature of the food manufacturing industry has changed. Food businesses are essential for life and have been for thousands of years. Now, more than six hundred million family farmers working on plots of land that are almost always less than twenty hectares (fifty acres) in size produce 80 percent of the food in the world, much of it consumed locally. These family farmers are, as stated by the UN Food and Agriculture Organization, "key drivers of sustainable development, including ending hunger and all forms of malnutrition."[10] Using the foods produced by this huge base of family farmers, countless types of suitable food systems and supplies—and thus a great variety of dietary patterns—have been created, developed, and sustained.

But now, food producing, manufacturing, retailing, and catering businesses throughout the world are being displaced by a relatively small number of colossal transnational corporations dependent on the manufacture or distribution of ultra-processed foods. These corporations have proliferated since the 1980s using the foreign direct investment system that enables them to create or take over businesses in any country.[11] The individual annual revenue of the largest corporations is as high as the annual gross domestic product of middle-size countries, and, unlike many national governments, these companies are able to plan strategically and divert or invest billions in new technologies and markets.[12] They have created a monolithic, global corporate food system.

It is sometimes claimed or implied that, as populations increase, world food and nutrition security will depend on a single dominant global food system, and thus on the types of food manufactured and sold by corporations. In fact, it is national, regional, and local businesses that need recognition, protection, and development.

ULTRA-PROCESSED FOODS ARE AN ABERRATION

The Nova classification's detailed description and listing of ultra-processed foods has enabled accumulation of comparable studies of their effects on diets and health. Since 2014, hundreds of studies on the effects of ultra-processed foods, conducted on all continents, have been published in scientific journals. Systematic reviews and meta-analyses of these studies show conclusively that these foods damage the quality of diets, increase the incidence of a large number of disorders and diseases, and reduce life expectancy.

Degradation of Diets
A meta-analysis of data from national studies in thirteen middle- and high-income countries, with average consumption of ultra-processed food ranging from 15.9 percent of dietary energy (Colombia) to 57.5 percent (United States), showed that increased ultra-processed food consumption

determines higher dietary free sugars and saturated fats and lower fiber, protein, potassium, and several micronutrients, as well as increased total energy intake.[13] The predicted mean content of free sugars of diets with 15 percent, 50 percent, and 75 percent of total energy from ultra-processed foods was respectively 9.6 percent, 15.3 percent, and 19.4 percent. For the dietary content of fiber, it was respectively 13.1, 10.7, and 9.0 grams per one thousand calories. Also, the higher the dietary share of ultra-processed foods, the lower the share of unprocessed or minimally processed foods such as vegetables, legumes, and fruits.

Analysis of nationally representative samples of children and adolescents from eight countries found that increased dietary share of ultra-processed foods implies higher dietary energy density and free sugars and lower fiber, liable to cause obesity.[14]

Analysis of US official surveys showed that the higher the dietary share of ultra-processed foods, the further away diets were from the recommendations in the Dietary Guidelines for Americans.[15] Across quintiles of ultra-processed food consumption, the percentage of participants with poor-quality diets more than doubled in children (from 31.3 percent to 71.6 percent) and more than tripled in adults (from 18.1 percent to 59.7 percent).

Harm to Health

More than thirty cohort studies have shown that high ultra-processed food consumption leads to ill-health.[16] Two were conducted in nine European countries and in nineteen countries of average income levels. Others were conducted in Brazil, China, France, Italy, Mexico, Spain, the United Kingdom, and the United States.

Disorders and diseases evidently associated with increased dietary share of ultra-processed foods include overweight, obesity, abdominal obesity, visceral adiposity, increased adiposity from childhood to early adulthood, type 2 diabetes, hyperuricemia, hypertension, cerebrovascular disease, dyslipidemias, coronary heart disease, breast cancer, nonalcoholic liver disease, renal function decline, Crohn's disease, frailty, depression, and cardiovascular, cerebrovascular, and all-cause mortality.[17] Diets high in ultra-processed foods also have been shown to increase the severity of

COVID-19.[18] Whether they have a similar effect on other infectious diseases is not yet known.

Meta-analyses of data on health outcomes for which there was more than one cohort study, most of high quality, show increased risk of overweight and obesity, type 2 diabetes, depression, cardiovascular and cerebrovascular disease and death, and all-cause mortality.[19]

A short-term, randomized controlled trial conducted at the US National Institutes of Health shows that ultra-processed diets caused an increase in freely chosen energy intake and consequent weight gain. Over a two-week period, twenty young adults consuming a diet with around 83 percent of energy from ultra-processed foods consumed approximately five hundred more calories a day than when consuming a diet with no ultra-processed foods. Participants gained 0.9 kilograms at the end of the two weeks with the ultra-processed diet and lost 0.9 kilograms, mostly body fat, by the end of the non-ultra-processed diet.[20]

These findings are consistent, graded, and plausible. No well-designed, independently conducted study contradicts them. Contrary opinions are usually expressed by commentators apparently ignorant of the literature or with conflicted interests or obstinate contrary views. It is now proved beyond reasonable doubt that food systems and supplies, dietary patterns, and personal diets high in ultra-processed products increase the risk of many more disorders and diseases than considered in orthodox dietary guidance. They may, in time, be shown to damage most if not all physical and mental systems and states. They are especially dangerous for children and young people, who tend to consume unusually high amounts of ultra-processed food and whose rates of obesity and diabetes in most countries continue to rise rapidly.

WHY ULTRA-PROCESSED FOODS ARE HARMFUL

There are plenty of reasons why ultra-processed foods are harmful, other than overall being high in fats, sugars, and salt and low in fiber. Many of these reasons are already demonstrated by investigations. Others are plausible observations or deductions. It is improbable that any specific reason is predominant and impossible that any one, if eliminated, would make ultra-processed foods innocuous. None would be resolved by reformulation to make ultra-processed foods contain less saturated fat, fat, sugar, or salt.[21]

Senior nutrition scientists are also commonly trained in medicine and epidemiology but are unlikely to have knowledge of other specialties, such as food science and technology or toxicology. This may well be why many direct ill effects of ultra-processed foods are neglected, minimized, or overlooked in conventional dietary guidance. These impacts include the following.

Biochemical

Ultra-processed foods contain analogues of the chemicals in food originally identified as nutrients because they promote growth or protect against various deficiencies. Many combine forms of fat and sugar, a combination rarely if ever found in nature. They contain few or none of the myriad bioactive substances that are not identified as nutrients, especially the phytochemicals in plant foods such as terpenoids, polyphenols, phenolic constituents, alkaloids, and saponins, absence of which may cause vulnerability to all sorts of disorders and diseases.[22]

Appetitive

The natural structure of whole food, including its fiber, is damaged or destroyed by ultra-processing. This is liable to derange or delay satiation and induce overeating. The same impact is produced by the hyper-delicious qualities that are engineered into many ultra-processed foods and debauch appetite.[23]

Metabolic and Microbiological

If the body does not recognize ultra-processed foods, with their xenobiotic foreign chemicals, as nourishing, then their metabolites may be shunted into adipose tissue, leaving consumers hungry. This would explain rapid development of obesity and gross obesity and other diseases in susceptible people.[24] Also, the naturally predominant beneficial bacterial species in the gut may be replaced by species that, by interaction with ultra-processed metabolites, are pathogenic.[25]

Toxicological

Most types of additives are contained in ultra-processed foods; indeed, several thousand additives are used in food.[26] The significance of additives is generally overlooked or ignored in orthodox dietary guidance. They are not mentioned in the 2020–2025 DGAs.[27] But they are not inert. A French study estimates that overall consumption of additives averages around 4 kilograms (8.5 pounds) a year, and that people who consume a lot of ultra-processed food may average up to 10 kilograms (22 pounds) a year.[28] Many if not most ultra-processed foods include flavors not individually identified on nutrition labels, formulated by companies that also create perfumes and scents for other products.[29] Some emulsifiers, dyes and other colors, flavor enhancers, and non-sugar sweeteners have been identified as causing ill-health.[30] Compounds generated during ultra-processing—for example, acrylamide and acrolein—increase oxidative stress and inflammation. Chemicals released from packaging materials, such as bisphenol A and phthalates, disrupt the endocrine system, whose functions include coordination of metabolism, response to stress and injury, and mood.[31] The effects of "cocktails" of additives are not officially checked and in general are unknown.

Addictive

Many ultra-processed foods are deliberately formulated to be hyper-palatable, instantly gratifying, and habit-forming.[32] A recent review of the science concludes that some "are created in ways that parallel the development of addictive drugs, including the inclusion of an unnaturally high dose of rewarding ingredients that are rapidly absorbed into the system and enhanced through additives." Furthermore, "when addictive substances become cheap, easily accessible, heavily marketed, and socially acceptable to use, the prevalence of addictive responses will increase."[33]

Technological

Humans are evolved to consume foods found in nature, developed and modified over many generations by agriculture, industry, and experiment and shaped by prevailing resources, climate, terrain, and custom.[34] It is a bold and unsupported assumption that humans can fully adapt to foods made from synthetic substances now constantly being invented by sophisticated technology. The body may well not recognize ultra-processed foods as nourishing but see them as useless, harmful, or toxic. Thus, depending on the amounts of these foods consumed, the body may become obese and weakened or diseased in various ways, according to which system or organ is most vulnerable.

Familial, Social, Cultural, Industrial, Environmental, and Corporate
In addition to the personal impacts, there are a range of broader ill-effects of ultra-processed foods.

Familial and social. Ultra-processed foods discourage preparation of meals, eliminate regular times for eating, disrupt family life, and frustrate conviviality. Ultra-processed foods are available in all sorts of outlets every day, at all hours, and are consumed anytime, anywhere, usually as snacks—from large supermarkets to filling stations, restaurants to kiosks— or sometimes as quasi-meals sold in "drive-ins" to be consumed while traveling in cars. Many are aggressively marketed, especially to young people, and are often cheap and sold in large or super sizes. All this induces overeating.[35] The habit of eating ready-to-heat or ready-to-eat ultra-processed snacks alone is surely one reason for what has in many countries become a pervasive sense of alienation and isolation. Societies erode when food generally ceases to be consumed as meals.

Cultural. The worldwide campaigns of transnational food manufacturing and catering corporations to promote their products are liable to make young people and adults ignore, forget, or reject the culture, traditions, and identity of their own country, region, and ethnicity.[36] This effect is most potent in the southern countries of Asia, Africa, and Latin America. It is likely to be less evident in the United States and other countries where individualism and consumerism have become dominant.

Industrial. As stated above, established, authentic national and local food supplies and dietary patterns depend on around six hundred million family farmers now producing 80 percent of the world's food, and on countless local, regional, and national manufacturers, retailers, caterers, and other industries.[37] But they are gradually becoming controlled, marginalized, purchased, or bankrupted by the corporations responsible for ultra-processed food.

Environmental. Production of ultra-processed food overuses and abuses land, water, and energy, contributing to climate disruption and creating pollution and waste.[38]

Corporate. Giant transnational corporations make most ultra-processed foods, whose ingredients are cheap and which are very profitable. The corporations are resourced by vast farms that feed and rear animals intensively or else grow a few crops, such as high-yielding strains of corn, soy, wheat, and other sources of oils, starches, and sugars. They have colossal budgets for product development, publicity, and pressure. While these companies are in competition with one another for similar products—such as soft drinks, breakfast "cereals," and burgers—they have interests in common and, as evident by their representative organizations and publicity companies, work as teams. Many have the declared intention to achieve annual double-digit (10 percent or more) sales growth in lower-income countries. They fund defensive research and academic departments of nutrition in some universities, support susceptible politicians, and seek to influence policies of UN agencies and national governments. Their evident overall ambition is to create a global, monolithic corporate food system that dominates discourse, controls food supplies, and shapes food patterns throughout the world.[39] This is perhaps less obvious in the United States, where what transnational corporations have already achieved may seem normal.

THE GOOD FOOD NEWS

So that is the news about bad food. What, then, is the good food news? What should be eaten, not only to protect against disorders and diseases but also to promote good health and well-being and the enjoyment of personal, family, social, and professional life, now and in the future?

These are not new questions. In the ancient Greek world, addressing them was one purpose of *diaita* (δίαιτᾰ)—usually translated as "dietetics"—the natural philosophy that actually means "way of life" or "way of being," which was displaced in the nineteenth century by the science of nutrition.[40]

The Brazilian guidelines are based on five explicit principles.[41] One is "Healthy diets derive from socially and environmentally sustainable food systems," meaning those that are created by and support the livelihoods of farmers and traders and that conserve soil, water, energy, forests, and other resources. For example, in Brazil, by law, one-third of the food supplied to state schools for the free meals eaten by forty-five million children must come from local smallholder farms, of which there are five million in Brazil.[42]

The four recommendations of the Brazilian guidelines are derived from the five principles. They apply universally and are as follows. The first is positive: "Make a variety of natural or minimally processed foods, mostly plants, the basis of your diet." The next two are cautious: "Use oils, fats, salt and sugar in small amounts for seasoning and cooking foods and to create culinary preparations," and "Limit the use of processed foods, consuming them in small amounts as ingredients in culinary preparations or as part of meals based on natural or minimally processed foods." The fourth is a warning: "Avoid ultra-processed foods."

Similar recommendations are now made in the dietary guides of various Latin American countries and of Israel and Malaysia.[43] The French High Council of Public Health includes a goal of a 20 percent reduction in ultra-processed food consumption.[44] "Choose minimally processed foods instead of ultra-processed foods" is a 2021 recommendation of the American Heart Association.[45] In 2021, the European Association for the Study of the Liver–*Lancet* Commission made similar recommendations.[46]

The overall positive "golden rule" of the Brazilian guidelines is "Always prefer unprocessed or minimally processed foods and freshly made meals to ultra-processed foods." This rule is also included in the 2021 Dietary Guidelines for Brazilian Children Under Two Years of Age.[47]

The guidelines include twenty-four pictures of meals, modeled on what roughly one-fifth of the Brazilian population actually eats. These habitual diets are mainly based on unprocessed and minimally processed food, prepared and cooked into fresh meals, combined with processed culinary ingredients and small amounts of processed food. They contain almost no energy in the form of ultra-processed food. They are close to the World Health Organization goals for most nutrients.

Latin America, like some parts of continental Europe, largely remains a collective community- and family-based culture, within which shared meals remain a central part of daily life. The first dietary guidelines prepared for Latin America address people not as individuals but as group members, on the grounds that "the family eats from the same pot."[48] The initial and current Brazilian official national dietary guidelines are also addressed to community and family members and do not refer to people as individuals.[49]

People and meals are social. Freshly prepared and cooked meals are not merely a substantial amount of food and drink consumed on one occasion. In most societies, they normally are family or social occasions.

The common custom of breakfast, lunch, and dinner (or supper), typically consumed with other family or household members, is one pattern of mealtimes. What is consumed at daily meals, such as breakfast, or at more substantial meals—such as weekend lunches and special occasions with invited family, friends, colleagues, or guests—usually has common features.

Commensality, the sharing of meals, is a defining characteristic of human society. When people eat fresh meals by themselves, there is still likely to be a sense of sharing, as felt within a family home or even in a restaurant. Until very recently in history, it was most unusual for people anywhere not to eat meals, or to eat in isolation from one another.[50]

Improvement of the nourishment of any population—meaning promotion of good health and well-being, as well as protection against illness—requires appreciation, preservation, and restoration of fresh meals. The development of human civilization always involves shared meals. In this respect, humans are no different from animals in nature, whose nests or lairs are also the places in which the parents feed their young. Herd animals eat and drink together, so the group is protected and the young learn. Lions and wolves hunt and eat together. Bees and monkeys gather material for food, or food itself, and share it.

This is also human nature. Remains of ancient human settlements show communities centered on dwellings where meals were prepared and consumed. Claude Lévi-Strauss points out that roasting, which followed the control of fire, is characteristic of societies where food is hunted, whereas boiling water requires a pot. "The roasted is on the side of nature, the boiled is on the side of culture."[51]

The classic stews of the world, originally prepared just for special shared occasions, are usually based on meat or fish, with varieties of legumes, grains, roots, vegetables, and herbs. Examples are the French cassoulet and bouillabaisse and the Brazilian feijoada and *moqueca*. They have been developed, refined, and adapted over many generations, cooked in a pot, with other pots for accompanying dishes, around which the family gathers. These traditions promote social health as well as healthy eating.

THE SIGNIFICANCE OF DIETARY PATTERNS

All over the world, vast repositories of knowledge have been assembled and used concerning sources of food, suitable methods of production, and how best to prepare and cook varieties of fresh and modified natural foods into delicious, sustaining, and nourishing daily meals. These are a central part of national and regional cultures and are an aspect of civilizations. In most countries, when food supplies are secure so that people regularly have enough to eat, everyday meals have normally been nourishing and satisfying.

This is how the habitual diets and cuisines of Mediterranean and Middle Eastern countries, of regions within India, China, and elsewhere in Asia, and of Mexico, Peru, Brazil, and other Latin American countries were created and have been developed and maintained over many centuries. The United States, Canada, the United Kingdom, and Australia are exceptional: while they have ethnic cuisines, local food customs, and gastronomic movements, no common long-standing national or regional dietary patterns are apparent.

Everyday, freshly prepared and cooked meals in societies with long-established staple food supplies and dietary patterns are usually based on combinations of available grains, roots and tubers, legumes, and vegetables. In Brazil, these are varieties of beans or sometimes lentils and rice, corn, or *mandioca* (cassava), with *couve* (cabbage) or other greens. Cuisines vary regionally according to what foods are grown and are influenced by various customs and cultures—Native, Black, or European. Herbs often and spices sometimes are used. A feature of Brazilian cooking is garlic, on sale in supermarkets next to onions in heaps. Animal foods are usually but not always included in main meals that may also include salads. Fruits are often eaten, especially at breakfast.

In Mexico and Central America, the staple combinations are similar: corn, beans, and squashes such as pumpkin and zucchini. African and Asian everyday meals are made by many methods and look and taste different, while also combining grains, legumes, roots or tubers, and vegetables.

"Blue Zones" are five areas of the world whose people express happiness, remain active, have very low levels of cancer and heart disease, and often live to great ages. The long-established dietary patterns of the Blue Zones of Nicoya in Costa Rica, Sardinia in Italy, Ikaria in Greece, and Okinawa in Japan are all based on grains, roots or tubers, beans, vegetables, and fruits, with a little meat, dairy produce, or fish. Their food is sourced locally and freshly prepared and cooked. The fifth zone, Loma Linda in the United States, is vegan.[52]

So, evolved dietary patterns are consistent. In most parts of the world, they have been and still often remain plant-based, mainly made from a mixture of grains, roots and tubers, beans and other legumes, leafy, salad, and root vegetables, and some meat and dairy produce, prepared and cooked with oils, often with herbs and sometimes with spices added, and fruits.

But in many parts of the world, dietary patterns now contain excessive meat or have too much sugar or salt added. All such diets deteriorate when they come to include ultra-processed food.

The various versions of the Mediterranean diet were developed over many centuries in sunny, temperate regions close to the sea, where olive trees and fish are abundant. They have been shaped by the climates, terrains, customs, and cultures of southern France, Italy, Greece, and North Africa.

These diets are identified as "intangible cultural heritage" by the UN Educational, Scientific, and Cultural Organization (UNESCO). They involve "skills, knowledge, rituals, symbols and traditions concerning crops, harvesting, fishing, animal husbandry, conservation, processing, cooking, and particularly the sharing and consumption of food." Eating together is "an affirmation and renewal of family, group or community identity. The Mediterranean diet emphasizes values of hospitality, neighborliness, intercultural dialogue and creativity, and a way of life guided by respect for diversity."[53]

In her book on Middle Eastern cooking, the food writer Claudia Roden, brought up in Egypt, evokes the familial, social, and cultural significance of that country's long-established meals.[54] "Friday night dinners at my parents, and gatherings of friends at my own home, have been opportunities to rejoice in our food.... Each dish has filled our house in turn with the smells of the *Muski*, the Cairo market.... They have conjured up memories of street vendors, bakeries and pastry shops, and of the brilliant colours and sounds of the markets."

It is often said that the Mediterranean diet is ideal. This misunderstands the nature of dietary patterns. Mediterranean dietary patterns, where they survive, are suited to their part of the world. They can inspire and valorize other dietary patterns in countries with their own histories, beliefs, resources, and methods of farming and cooking.

Some of these also have ancient food systems and dietary patterns. For example, in the city of Hangzhou in central China, according to Marco Polo in the thirteenth century CE, "markets were 'innumerable,'" selling game, varieties of apricot, early eggplants, and live baby fish carried inland in baskets. Philosophers described the principles of cooking and diet. Eminent officials toured the surrounding countryside looking for "natural" cooking, for dishes that preserved "the food's basic nature." The past was interpreted through food.[55] The region of Zhejiang, which includes Hangzhou, has revived and renewed such traditions. Here, too, everyday food is plant-based.[56]

By sharing meals, companions, families, friends, guests, travelers, colleagues, and members of clubs, associations, institutions, and societies as a whole come together, get to know one another, and become and remain coherent. Family agreements and professional and political decisions at all levels are made after breaking bread together. This has always been so.

WHAT IS TO BE DONE?

So the best food advice for people personally and as partners, family members, customers, and consumers is clear. For good health and well-being, eat freshly prepared and cooked meals, mostly based on a variety of plant foods, preferably enjoyed in company, and avoid ultra-processed food.

However, personal guidelines are necessary but not sufficient. There are limits to what people and families can do, especially when they lack resources. Public policies and actions are needed in order to encourage freshly prepared and cooked meals and correspondingly to discourage ultra-processed foods. Public health policies and actions, including statutory and fiscal regulations, are most effective when they are concerted, positive, transparent, and equitable.

What follows are some proposals, including a number already published, for those who shape dietary patterns, food supplies, and food systems, including UN agencies, national governments, industries, civil society organizations, social movements, educators, and those responsible for institutional meals.[57] The proposals are local, regional, national, or international. All need prior consultation, full explanation, and wide publicity in social, broadcast, and print media.

A report on best food practices, to specify the actions below and others as identified and agreed, needs to be compiled, constantly updated, and published on the websites of all relevant UN agencies and national government departments. The task can be outsourced to a competent professional or academic organization, to be recognized as a UN collaborating center.

More Freshly Prepared and Cooked Plant-Based Meals

Governments and all relevant government departments, the United Nations and all relevant UN agencies, and health and food authorities at all levels, to review and revise their statements and guides on diet, so that these emphasize freshly prepared and cooked plant-based meals, mainly made of unprocessed or minimally processed foods, as well as avoidance of ultra-processed foods.

Governments to estimate and report on the personal, family, social, cultural, economic, and environmental costs and benefits of plant-based dietary patterns nationally and globally. The reports to be widely publicized and updated regularly.

Bodies that determine international, national, municipal, and local food policies to include representatives of citizens, small and family businesses, and Indigenous peoples, as well as scientists, professionals, and officials.

Manufacturers to be encouraged to create, maintain, develop, or improve minimal processing methods that protect whole foods, preserve their taste and flavor, and make their preparation and cooking as fresh meals easier and more varied.

Retailers to display fresh fruits near checkout counters, stop displaying vegetables and fruits only in unblemished form, and sell bruised or damaged produce at lower prices.

Farmers and their workers who produce plant foods sold for consumption in whole form to be granted direct and indirect subsidies, including income support and tax benefits, and their contribution to food and nutrition security to be recognized and valorized.

Farmers' markets, street traders, public cafeterias, and restaurants that serve freshly prepared meals made according to local, regional, or national customs to be established or encouraged, and subsidized where appropriate, in main locations of urban areas.

School, hospital, prison, and cafeteria meals to be plant-based and freshly prepared and cooked. State school lunches to be free.

Elementary and junior high schools to specify household economics, including meal preparation and cooking skills, in their core curricula; to be an option at the senior-high level.

Locations that preserve and sustain healthy dietary patterns to be identified and given national, well-publicized annual awards.

Less Ultra-Processed Food

The positive actions above can be accompanied or followed by actions designed to reduce production and consumption of ultra-processed foods. These will be resisted by many transnational corporations whose profits depend on these foods but are likely to be welcomed by most other food businesses.

Public policies and actions here can be guided by the experience of tobacco control, bearing in mind that, while food is essential, there is no need to consume ultra-processed food. As with tobacco, actions need to involve statutory measures, including fiscal regulations.

The personal, family, social, cultural, economic, and environmental costs of ultra-processed food, nationally and globally, including on health services, to be estimated, reported, widely publicized, and updated regularly.

Bodies that determine international, national, municipal, and local food policies, to exclude as members, observers, or advisers people whose commercial or professional interests conflict with those of public health.

The most effective way to reduce ultra-processed foods would be to prohibit all classes of additives now used exclusively in their manufacture. This needs to be known, and some governments might favor the policy, but it probably would not gain public support. Short of this, colors and flavors could be prohibited, generally or in specified commonly consumed products.

Governments to tax ultra-processed products and subsume existing taxes on sugared ultra-processed food into this general tax.

Governments to require all ultra-processed foods to be identified as such on labels and promotional materials, with warning messages to be specified by national ministries of health. Front labels to also exclude misleading visual and verbal claims and descriptions implying that the product contained is fresh.

Manufacturers to cease reformulating foods that remain ultra-processed and instead make minimally processed foods, processed culinary ingredients, and processed foods (which prohibition of specified additives would encourage) and diversify into making other products that are harmless or beneficial to health.

Businesses that grow or make ingredients for ultra-processed foods or that manufacture, distribute, or sell ultra-processed foods to have any direct and indirect subsidies—including income support and tax benefits granted for these purposes—removed.

Retailers to cease making price-discounted offers on ultra-processed foods.

Schools and hospitals to be designated as healthy food zones, and the sale of ultra-processed foods in and within a generally agreed distance of schools and hospitals to be prohibited.

School, hospital, and prison meals to exclude ultra-processed food. Cafeterias for workers and armed forces to limit it to under a generally agreed maximum.

Schools to teach how to manage the media, including studying the propaganda designed to sell ultra-processed foods to children and young people.

A FINAL WORD

The pathologist and politician Rudolf Virchow, a founder of social medicine, said in 1848, "Epidemics are great warning signs, against which the progress of civilization can be judged." The pandemics of obesity and diabetes are not the result of a global collapse of personal willpower. Their specific cause is the rapid rise, especially since the 1980s, in the production, dissemination, and consumption of ultra-processed foods, which are chemical concoctions disguised as hyper-delicious real food. Their immediate context is the freedom given to transnational food corporations that make vast profits from ultra-processed foods, whose ingredients are uniform and cheap, to invade countries and subvert long-established rational, authentic food systems and dietary patterns.

Most readers of this chapter and their families can choose to avoid ultra-processed foods. But this will not stop their increase, or the increase in obesity and diabetes now most evident in lower-income countries. Further, policies and actions specifically designed to reduce ultra-processed foods, including statutory and fiscal measures such as those listed above, no matter how widely adopted, can have only limited effect. This is because global development is seen as ever-increasing money, and the prevailing ideology in powerful countries is individualism. Dominant civilizations are progressing in a calamitous direction. Resources are squandered, nature is neglected, the living and physical world is trashed, climate is disrupted, corporations become bloated, billionaires subvert countries, family and community life is denigrated, future generations are ignored, history is forgotten, drug abuse is rampant, and rates of physical, mental, and emotional disorders and diseases rise. Ultra-processed food is an example and a part of what has become a global syndrome.

This chapter advocates freshly prepared, commensal meals. This implies taking care of ourselves with our families, friends, and colleagues, and taking time to acquire, prepare, cook, and enjoy meals made with fresh food. These are acts of affection and, in societies where long-standing food patterns survive, part of local or national culture. It is often claimed that, these days, there is no time to make meals; how absurd, in countries where six or so hours a day are spent watching television or playing video games. Elimination of ultra-processed foods depends on development being seen as ever-increasing well-being and humans as social beings within nature, with all this implies. *Bom apetite*, as is said in Brazil. Good appetite!

4.

Dairy Together: Fighting for a System That Gives Small Farmers a Fair Shake*

By Sarah E. Lloyd

Sarah E. Lloyd dairy farms with her husband and his family on the 450-cow Nelson family farm outside Wisconsin Dells. She works off-farm as a supply-chain development specialist for the University of Minnesota on the Grassland 2.0 project, looking at building production and markets for regenerative agricultural systems. She also works on development for the Wisconsin Food Hub Cooperative, a fresh produce business owned by the produce farmers and the Wisconsin Farmers Union. She is on the board of the Wisconsin Farmers Union and the Wormfarm Institute and has served on the National Dairy Board and the Wisconsin Milk Marketing Board. Lloyd has a PhD in rural sociology from University of Wisconsin, Madison, and a master's in rural development from the Swedish University of Agricultural Sciences.

---

* Portions of this essay were originally published on the Disparity to Parity website, March 31, 2021, at https://disparitytoparity.org/dairy-together-building-a-farmer-led-movement-for-supply-management/. Reprinted by permission of the author.

We US dairy farmers have been on a roller coaster for a long time. Volatile prices going from highs to lows on a two- or three-year cycle have been pretty standard going back to the 1990s, if not before. But the years since around 2010 have seen a change in the intensity of this volatility. Prices drop really, really low—well below what it costs for us to produce the milk we ship on the milk truck every day. Then prices shoot up dizzyingly high—high enough to get us excited when we get the milk check on the seventeenth of the month. We think, "Okay, we can do this! Think of how much of our debt we can pay off if the prices stay this high. We can finally replace that twenty-year-old piece of machinery that breaks down all the time." But then prices plummet a couple of months later, just when we thought we'd be able to get our feet under us, and our spirits plummet with them.

Take 2020, for example. The Class III is the price paid to farmers for milk going into cheese processing. In Wisconsin, the "cheesehead" state, that's where most of the milk goes, including the milk from my own family farm. During 2020, the Class III for one hundred pounds of milk swung from $12 to almost $25. How are we supposed to deal with that kind of unpredictability?

You might say, "Well, just grab onto the highs and soldier on through the lows." That's basically what we dairy farmers try to do. Unfortunately, the lows seem to last longer than the highs. Meanwhile, the cost of all the inputs—fuel, seed, fertilizer, supplies, labor—keeps going up, and the lows aren't getting any higher to make up for them. We call it the cost-price squeeze, and the result is a treacherous business landscape for farms, especially small and midsize ones like ours. Such farms used to dot the midwestern landscape by the thousands. They were economic and social engines that bought things from other small and midsize businesses, like the local grocery store and hardware store, and supported the neighborhood schools, churches, and other social spaces. Now these farms are dying. In Wisconsin, the number of dairy farms has dropped from over fourteen thousand in 2007 to just over six thousand in 2023.

Economic realities in the United States have left dairy farmers with two choices: get big or get out. That's the challenge my husband, his brother, and my in-laws faced back in 2002. They expanded the herd from

125 cows to 250, trying to run a little faster on the treadmill, hoping to get more milk flowing out the farm gate and more money coming in for the three families trying to make a living from the farm.

This is the same path many other dairy farmers have chosen. Unfortunately, things don't get any easier, even when the treadmill starts spinning faster. To expand the herd, you borrow money from the bank to build some new facilities—in our case, a double-eight milking parlor where we can milk sixteen cows at a time, along with new barns to house the bigger herd. You also need to buy or rent more land to grow crops for the additional cows, and you need to hire more workers to help run the farm—all to keep your business running and to make your loan payments on time. With each step, the treadmill feels as if it's speeding up, and maybe the incline is rising, or it's one of those fancy treadmills where the resistance can get ratcheted up.

That's what we are doing here at our farm: running on the treadmill and trying to figure out how to jump off, or at least slow it down to a sane pace.

My husband, Nels Nelson, and I dairy farm with his family in Columbia County, Wisconsin, about an hour north of Madison, on the farm that has been in the Nelson family for over a hundred years. I married the farm—I mean, my husband—in 2007.* I didn't grow up on a farm, but until 1997 my father and his sister were part owners in their home dairy operation, where my father grew up, his dad grew up, and his grandfather had started farming with my great-grandmother back in the early 1900s. My dad left the dairy farm at eighteen to go to college. Dad and I now chuckle about the fact that I am dairy farming in his home county, just thirty-six miles west of the Lloyd ancestral farm. He jokingly shakes his fist, à la Grandpa Simpson, and wails, "Sarah, I tried to get us out of dairy farming, and here you are right back where I started. Why? Why? Why!?"

Today, the size of the Nelson family farm has crept up to 450 cows (can you feel the treadmill racing?). It feels gigantic, especially when compared to the twenty-five and fifty cows that the Nelson and Lloyd families milked here in Columbia County just fifty years ago. But in reality, our farm is just midsize; as of 2022, the average herd size in Wisconsin

* That's a joke. Love you, Nels.

was 204 cows.[1] There are now dairy farms in Wisconsin with ten thousand cows, and as you head west to the new dairy-producing areas of the United States—like South Dakota, New Mexico, and Idaho—you see fifteen-thousand-, twenty-thousand-, and even thirty-thousand-cow operations. This is why, even as we see rapid farm exit—people choosing to get out of dairy farming, jumping off the treadmill or sadly declaring bankruptcy because of the cost-price squeeze—we are *not* reducing our production of milk in Wisconsin or in the country. In fact, as Figure 1 shows, milk production in the United States increased by 13 percent from 2012 to 2021.[2]

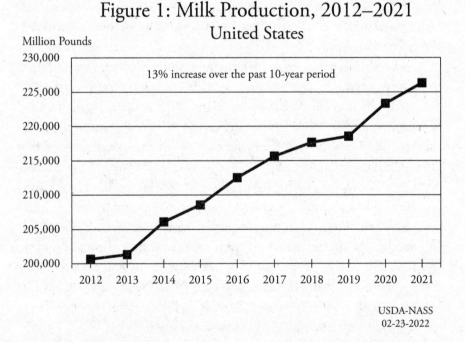

Figure 1: Milk Production, 2012–2021
United States

However, the cost-price squeeze impacts the big producers just as much as the small ones. And the ever-accelerating treadmill has created a strange situation in which dairy farmers are increasing production not only when prices are high but also when prices are low.

The first half of this equation makes sense (I took Econ 101 in college). The market sends a signal with a higher price, so you ratchet up production to meet that higher price signal and make more money. But why would you increase production when prices are low and you are losing money on every pound of milk you sell? It's because you owe the bank a bunch of money and you can't just turn the cow off. So you keep going and try to run just a bit faster, hoping to squeak by until you get back to a higher price period. But year after year, the vicious cycle—too much milk leading to low prices, creating a situation in which only the biggest farms can survive, which then leads to more milk produced and even lower prices—continues to speed up.

This can only spell long-term disaster for the families and communities that depend on dairy farming for their livelihood. The same trend toward fewer, larger farms also has serious ecological implications. Concentrating more and more animals in fewer places concentrates the ecological risk. Challenges like manure management and nutrient loading for giant dairy facilities, when not properly managed, pose severe risks to both groundwater and surface water quality, as well as other environmental problems.

PLAYING THE GAME OF MONOPOLY
AND MONOPSONY

The vicious cycle and cost-price squeeze that is driving farmers out of business and drying up our rural communities is not happening all by itself. The increasing power of a handful of corporate agribusiness and food system players is also a huge factor.[3] Dairy farmers suffer from the cascading effects of the increased market power of the companies we buy our supplies from—which economists refer to as monopoly power—as well as the increased market power of the companies we sell our milk to—that is, monopsony power.

Here's an example of how it works. At our farm, we use artificial insemination to breed the next generation of cows. In other words, rather than keeping bulls around to do our breeding, we have a breeder who comes every day and breeds our cows using commercially supplied semen. We keep all of the female young, known as heifer calves, and raise them up to be milking cows. The great majority of dairy farms follow the same practice.

What's the problem? Over the years, the industry of bull semen has become increasingly consolidated, like almost every other business in the agricultural sector. I saw the results several years ago when my husband came in the house from the barn looking upset.

"The price we have to pay for bull semen is going up," he told me. "Way up."

"Why is that?" I asked.

Nels recounted what the sales rep had told him. "We have to go after some of these bigger farms," the sales rep had said, "like that one a couple counties over. They're expanding their herd to eight thousand cows with a goal of being over ten thousand cows in a couple years. We need to offer them a volume discount. Otherwise we are going to lose their business. They have us over a barrel. So, to try to keep our company afloat, we need to raise your price."

This story shows how it works. A shrinking number of buyers and sellers gives a small group of big operators virtual control of the market, which means the little guy inevitably gets crushed.

The same dynamic happens in other parts of our business. We used to ship our milk to a cooperative cheese processor owned by a consortium of dairy farmers. This processor was supplying a lot of the mozzarella cheese used by a frozen pizza company. And almost every year, at the annual meeting of the smaller dairy owners, the manager of the processing plant would say, "The frozen pizza company is expanding to get bigger so they can keep up with the consolidating grocery sector. Walmart is taking a bigger and bigger share of the food business, and we need to be able to compete to get our products into their stores. To do that, we need to up our production. That is why we have to go after the biggest dairy farms, because we need to get more milk into those plants on a daily basis. Oh—and those giant farms are going to get a volume discount, and we need to take that out of somewhere. So the price we can pay to you is going to go down."

As you can see, concentration up and down the food chain keeps making things harder for small operations like our family dairy farm. In the retail grocery arena, Walmart is the biggest and most powerful. But the Kroger-Albertsons merger announced in 2022 has created a new mega-actor biting at Walmart's heels.[4]

The giant companies that now dominate American food production like to claim that their size and power are responsible for providing US citizens with cheap, abundant food. It's a misleading rhetorical tactic designed to pit farmers against consumers. In reality, the concentration of economic power in a few hands mainly has the effect of steadily increasing the profits enjoyed by the biggest corporations, at the expense of both farmers and consumers.

As food price spikes during the COVID-19 pandemic have shown, the prices that consumers pay are not closely related to the income farmers receive. Instead, the market control enjoyed by the monopolistic retail actors allows them to push prices up and then blame inflation, all while enjoying massive profits.

Idaho-based Albertsons Companies—which owns Safeway, Acme, and other grocery chains, with a total of 2,278 stores—reported a whopping $1.16 billion profit for the first three quarters of 2021, an increase from $994 million for the same period in 2020 and $399 million in 2019. Similarly, Cincinnati-based Kroger, which manages 2,750 stores nationwide, posted an operating profit of $2.5 billion in 2021 compared to $1.7 billion in 2019—a 47 percent increase during the pandemic. Big profit increases were also reported by other major corporations in the US food supply system, from processing firms like Conagra Brands to restaurant chains like Chipotle.

In the words of Kyle Herrig, the president of advocacy organization Accountable.US, "Corporations did not need to raise their prices so high on struggling families....Corporations have used inflation, the pandemic, and supply chain challenges as an excuse to exaggerate their own costs and then nickel and dime consumers."[5]

THE PRODUCTION TREADMILL AND THE FALSE SOLUTION OF EXPORTS

Looking back at history, I can understand how the current tendency toward chasing increased dairy production came about. It started as an honorable goal. During World War II, US farmers were asked to produce more to support the domestic population and pay for war bonds. After victory was won, farmers were encouraged to keep it up to feed a war-torn Europe.

Researching this period, I came across a notice on the front page of the Richland County, Wisconsin, *Republican Observer* for October 23, 1941, calling all farmers to a meeting about increasing production for the war effort (Figure 2). It read, in part: "The dairy farmers of Wisconsin are being asked to increase their milk production in 1942 over this year by an average of 400 pounds a cow. The hog production of the state is to be stepped-up by adding one sow for every four or five farms producing pork. And the hens of the state are expected to lay 11 dozen eggs for every 10 dozen they produced this year. Farmers should keep in mind and attend the meetings to be held in each Richland county school house on Friday evening, October 24th."

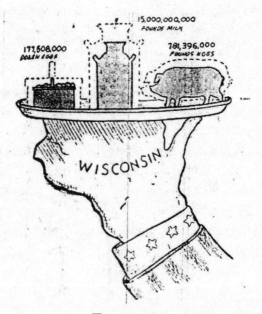

Figure 2

Wisconsin dairy farmers collectively were directed to increase production to fifteen billion pounds of milk. In subsequent decades, this expanded to include US farmers being asked to "feed the world."

We were responding first to wartime needs, then to severe famine in many parts of the world. I recently came across the letter and certificate that my maternal great-grandmother's family received in 1943 at their farm in Isanti County, Minnesota, "in recognition of your patriotic cooperation in the Food for Freedom program" (Figures 3 and 4).

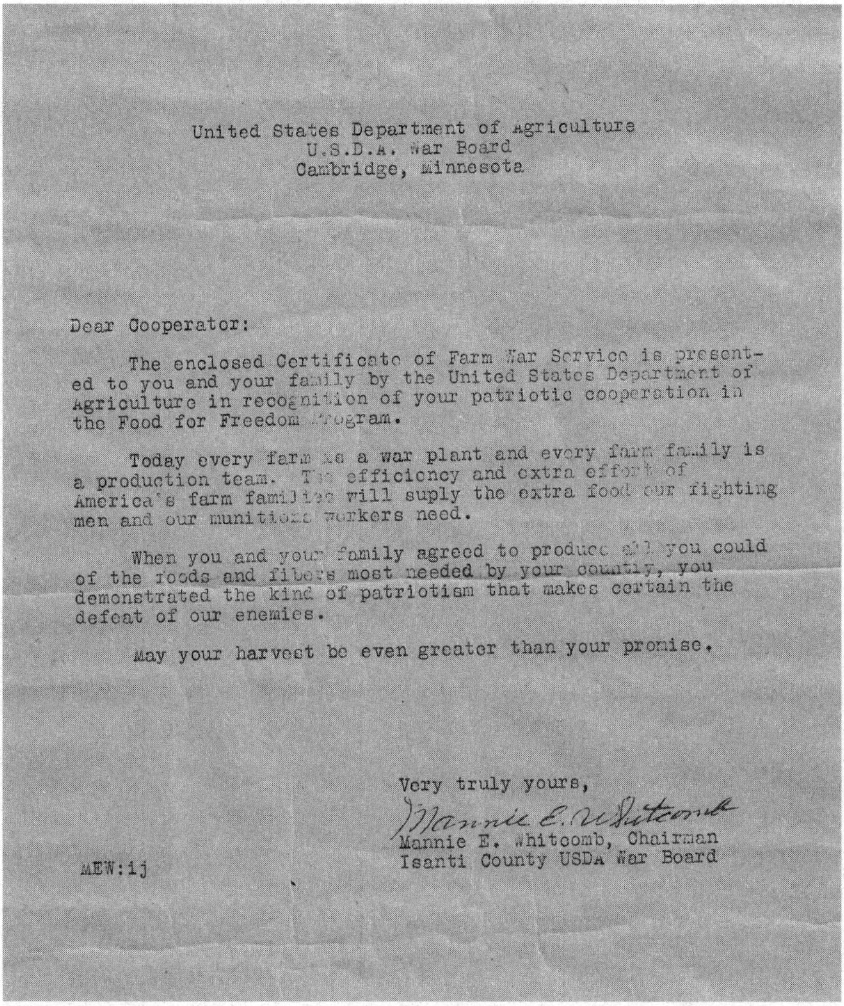

United States Department of Agriculture
U.S.D.A. War Board
Cambridge, Minnesota

Dear Cooperator:

The enclosed Certificate of Farm War Service is presented to you and your family by the United States Department of Agriculture in recognition of your patriotic cooperation in the Food for Freedom Program.

Today every farm is a war plant and every farm family is a production team. The efficiency and extra effort of America's farm families will suply the extra food our fighting men and our munitions workers need.

When you and your family agreed to produce all you could of the foods and fibers most needed by your country, you demonstrated the kind of patriotism that makes certain the defeat of our enemies.

May your harvest be even greater than your promise.

Very truly yours,

Mannie E. Whitcomb, Chairman
Isanti County USDA War Board

MEW:ij

Figure 3

US dairy farmers were proud to do our part in a time of emergency. But current research shows that the world no longer has a shortage of food. Instead, we have a distribution problem. We also have industrial agribusiness interests pushing to keep up ever-expanding exports of commodity food products like corn, soy, and milk powder, as well as of the technology and

Figure 4

inputs that farmers in other countries need to get on this same export tread-mill. (There are a number of great sources you can explore to learn more about how farmers both here and around the world get sucked into the vortex of geopolitics run by corporate profiteers. I'd start with Raj Patel's *Stuffed and Starved: The Hidden Battle for the World Food System* and Timothy A. Wise's *Eating Tomorrow: Agribusiness, Family Farmers, and the Battle for the Future of Food*.) We have been wildly successful in achieving the goal of expanded production. Technology and economic efficiencies gained with consolidation at all steps in the process have been cheerfully implemented by those taking the proceeds to the bank: the input suppliers, technology sellers, food processors, and grocery retailers. But farmers are struggling to make a living. When they go to the bank, it's to make a loan payment or to ask the bank to lend them more money so they can try to milk some more cows, just to squeak by.

Our dairy industry and political leaders present us with their silver-bullet solution for this problem: exports. "Don't worry. Keep on producing. We'll just export it!" And again, dairy farmers have done as they've been asked. We have reached the astounding outcome of increasing dairy exports from just 3.6 percent of total US production in 1996 to 15 percent in 2020—all while increasing overall production year over year.[6]

This increase is touted as a great accomplishment by dairy industry and government leaders at both the state and federal level—people like Tom Vilsack, secretary of agriculture under both Barack Obama and Joe Biden and the head of the US Dairy Export Council during the Trump administration. (The council, by the way, is overseen by the same USDA that Vilsack has twice led, a great example of the revolving-door politics of agriculture policy in Washington.)

But even as exports soar, we see a precipitous drop in the number of dairy farms. If chasing the export markets is such a winning strategy for small and midsize farms, why are they disappearing so quickly?

I am not an isolationist. I fully understand that the United States is involved in global markets. And I accept the argument that, to enjoy the profit from the high-value butterfat used in products like butter and ice cream, we have to find a home for the other proteins and solid components. But exporting cannot be a strategy pursued in isolation.[7]

MORE FARMS, NOT LARGER FARMS

Aldo Leopold, the noted "land ethic" thinker, explored the ecological impacts of the fixation with production metrics in his best-known essay, "The Land Ethic," in *A Sand County Almanac*. In dairy, the metrics on which we are fixated are hundredweights (cwt)—a hundred pounds—which is the unit we use to sell our milk to the market. One of our cows on average produces about eighty pounds of milk a day. Our farm ships about 35,000 pounds (350 cwt) each day with the milk truck that comes down our drive to take our milk to the cheese processor.

Of course, this statistic is important to us. But when production statistics dominate our thinking, the results can be destructive, as Leopold wrote: "By and large, our present problem is one of attitudes and implements. We are remodeling the Alhambra with a steam-shovel, and we are proud of our yardage. We shall hardly relinquish the shovel, which after all has many good points, but we are in need of gentler and more objective criteria for its successful use."[8] The Alhambra, in the Andalusia region of Spain, was built over hundreds of years starting in the eleventh century and is considered one of the most magnificent achievements of Islamic architecture.

Leopold's message is an important one. The benefits and risks of dairy production are deeply connected with the cycle of nutrients. Too much of a good thing in one place creates an ecological bad. Along with the negative economic and social impacts on family farms and rural communities from this chase for more and more production, the concentration of dairy production into fewer and fewer, larger and larger farms also concentrates ecological risk.

Can America's need for dairy products be met by a larger number of smaller farms? Absolutely. Here are some concrete statistics to demonstrate the point.

In 2021, Wisconsin dairy farms produced over 31.7 billion pounds of milk. The average Wisconsin cow produced 24,884 pounds of milk. To maintain 31.7 billion pounds of annual production, we need approximately 1,267,000 cows.[9]

Using round numbers of number of farms, cows, and annual production per cow, the chart below illustrates the different sizes of farms that can produce 31.7 billion pounds of milk in Wisconsin.

Number of Farms	Average Herd Size
12,500	100 cows
2,500	500 cows
1,250	1,000 cows
250	5,000 cows

The goal of having many diverse farms producing dairy products rather than a handful of giant operations is eminently practical. It will lead to a healthier and more sustainable environment as well as a more equitable agricultural economy.

DAIRY TOGETHER: A RESPONSE TO THE CRISIS

The issue of continued overproduction in America's dairy industry has not yet been addressed, which means that farmers are chronically operating in a surplus market. A coordinated supply response is needed. Instead of trying to impact the demand side through efforts promoting consumer consumption and expanding exports, farmers are now coming together to organize on the supply side, focusing on the need to manage supply to balance with demand.

This is not a new concept. For example, cranberry growers and potato growers have used it over the years in the United States. And we see a variety of systems in other countries that provide supply-side solutions to the problems of agriculture economics. These include a coordinated system covering the Canadian dairy industry, which works by using quotas to match supply with demand. The challenge is applying a similar approach that will work in the context of the US dairy industry.

Toward this end, in 2017 the Wisconsin Farmers Union began a concerted effort to build a farmer-led movement to call for reforms, with the goal of supporting family farms with a fair price for their milk. This work is being done under the banner of Dairy Together, a coalition of farmers, farm organizations, allied consumer and environmental organizations, and supply-chain partners that aims to:

- engage dairy farmers in discussions and peer-to-peer information sharing about dairy policy and strategies to address financial stress and policy options;
- establish a farmer network to support shared communications and messaging about the emerging economic situation and potential strategies for short- and longer-term action;
- provide opportunities for farmers to build consensus and raise the visibility of their concerns within their dairy co-ops and various public forums; and
- identify and contract for research opportunities that support policy and organizing.

This work was officially launched in March 2018 with a set of meetings around Wisconsin, bringing farmers together to hear from their Canadian dairy farmer peers about the way the Canadian coordinated system works. The goal was not to replicate the Canadian system but to learn lessons from our neighbors to the north.[10] The meetings then expanded to other states, including California, Michigan, Minnesota, New York, Ohio, Pennsylvania, and Vermont. Farmers and advocates from Wisconsin and Michigan also took an August 2018 bus trip to Albany, New York, to meet with other farmers and advocates to talk dairy-price policy reform at a gathering convened by the Agri-Mark dairy cooperative. Over four hundred people attended this meeting, and a number of different policy mechanisms were discussed.[11]

Continued Dairy Together organizing efforts have brought together state Farmers Union chapters from around the country, the California Dairy Coalition, and the National Farmers Organization, which was a co-partner on the Dairy Together Road Show. In addition, Farm Aid sponsored the bus trip and featured the issue at the 2019 Farm Aid concert at Alpine Valley, Wisconsin. The conversation has also attracted participants from the Northeast Organic Farming Association, local chapters of the Farm Bureau, and the National Family Farm Coalition and some of its member organizations.

A group of farmers from these organizations gathered in Washington, DC, in the summer of 2022 and met with members of the House and Senate and their staffs to explain the need for bold policy action. Of course, any federal law or regulation will need to go through the official policymaking process.

POLICY GOALS AND OPTIONS

With facilitation by Wisconsin Farmers Union staff, Dairy Together convenes regular national calls of farmers and coalition partners to discuss possible policy mechanisms and strategy to push decision-makers in the industry and the government to adopt a supply-management system. The farmers and organizations involved in the Dairy Together movement have agreed that national dairy policy should:

- provide a fair price to farmers;
- reduce price volatility;
- address the problem of overproduction;
- discourage industry consolidation;

- allow new farmers to enter the market without a huge financial hurdle;
- be national in scope and mandatory, with no opt-out clauses;
- be coupled with short-term action to sustain farmers until longer-term structural change can be enacted;
- address import and export issues and lead to smart trade policy; and
- include meaningful farmer input in development, implementation, and governance.

A number of specific policy options are under consideration. For example, Dairy Together commissioned dairy economists Mark Stephenson (University of Wisconsin, Madison) and Chuck Nicholson (then at Cornell University, now at University of Wisconsin) to reexamine a policy model created during the 2012 Farm Bill discussions.[12] This model, which grew out of efforts spearheaded by the Holstein Association, includes a market access fee as a main policy driver. It does not include a Canadian quota system, but it does establish a base level of production per farm. Farmers wishing to expand beyond that base would pay fees for additional production units, which would be pooled by a federal agency and redistributed to the farmers who do not expand beyond their base. This model would create a disincentive to increased production and provide a higher price for those who refrain from adding to surplus production.

The market access fee mechanism is by no means the only policy that would achieve the goals of Dairy Together. However, it is the model that has been most examined, was included in past Farm Bill deliberations, and received bipartisan support, although it was not passed into law.

The Dairy Together coalition continues to bring additional partners to the table. It has not yet presented a final detailed policy proposal, although it has developed an interim set of recommendations, called the Dairy Revitalization Plan, which includes the market access fee concept.[13]

Another important policy shift that dairy farmers need is a return to active enforcement of antitrust laws, particularly in the agricultural sector, at both the federal and state levels. These rules could be used to curb the economic and political power of the giant corporations that increasingly dominate US agriculture. State attorneys general can do a lot on this front, particularly if they join together with other states' attorneys general to act. In 2023, the Wisconsin Farmers Union is meeting with state legislators and the governor to ask for an increase to the personnel line in the attorney general's budget to support increased antitrust action.

On the federal level, USDA secretary Tom Vilsack could use his position and its political and policymaking power to stand up for an agricultural system that will support small farmers, communities, and the land. A strong stand by Vilsack could go a long way toward bringing more people and organizations into the movement now being built by the farmer-led efforts of Dairy Together.

THE CHOICE IS OURS

Wisconsin and the rest of the United States can choose how we would like to structure the production of dairy. We can enact policies to support small and mid-scale farms and have a system in which many such farms can exist and thrive independently. We can enact a system of federal price policies that sends a strong signal to reduce overproduction. We can enforce antitrust laws already on the books to limit consolidation and concentration in our economic systems and to give people a fair shot in the market.

We can do all this if we choose—or we can do nothing. As the Chinese proverb says, "If you stay on the path you are on, you will get where you are going." If we choose to do nothing, we will have dairy production on very few farms with very large herds, along with all the economic, social, and environmental problems this structure creates.

The choice is ours. The time for action is now!

5.

Food and Land Justice

By Leah Penniman

Leah Penniman is a Black Kreyol farmer, author, mother, and food justice activist who has been tending the soil and organizing for an anti-racist food system for twenty-five years. She currently serves as founding co-executive director and farm director of Soul Fire Farm in Grafton, New York, a Black- and Brown-led project that works toward food and land justice. Her books are *Farming While Black: Soul Fire Farm's Practical Guide to Liberation on the Land* and *Black Earth Wisdom: Soulful Conversations with Black Environmentalists*. Find out more about Penniman's work at www.soulfirefarm.org and follow her @soulfirefarm on Facebook, Twitter, and Instagram.

As a mother, I know of no greater yearning than the sacred imperative to feed my children. When Emet was a newborn and Neshima was just two, my partner Jonah and I moved to the South End of Albany, New York, a neighborhood deemed a "food desert" by the federal government due to the paucity of grocery stores and high poverty rates. I prefer the term "food apartheid," which indicates that the vast swaths of this nation where people of color disproportionately suffer from hunger, diabetes, and heart disease are the human-created outcomes of systemic racism, not natural phenomena.

Jonah and I struggled to feed our kids fresh produce, not for lack of effort but because there were no accessible grocery stores, farmers' markets, or community garden plots in our area, and no public transportation. We applied for benefits under the Special Supplemental Nutrition Program for Women, Infants, and Children (WIC), which provides federal grants for food assistance, but when I attempted to redeem my check at the corner store to purchase milk and eggs, the customer behind me spat on my shoes for holding up the line. The only way we could get greens and tomatoes was to join a costly community-supported agriculture (CSA) program with a pickup location 2.2 miles from our home. Each week, we packed up the kids and made the long trek on foot.

There was a cruel irony to the fact that Jonah and I had been laboring as farmworkers since we were teenagers but now were struggling to obtain crops for our own table. But our problem was a common one. According to a 2017 Department of Agriculture report, almost forty million Americans live in food deserts where they cannot access or afford the life-giving foods that make us whole.[1]

But our background gave Jonah and me something relevant to offer. When our neighbors learned that we were seasoned growers, they asked, "Why not start a farm for us? A farm for the people?"

We took that challenge seriously. In 2006, we wed ourselves to eighty acres of eroded mountainside land in Mohican territory near Grafton, New York. We spent years healing the soil with cover crops and mulch, regenerating the forest, building a solar-powered, straw-bale home and education center by hand, and assembling a farm team.

Soul Fire Farm opened in 2010, driven by the collective desire of Black and Brown families to feed ourselves. We established a vegetable and egg delivery program that allows members to choose how much to pay, as well as a home gardens program that provides lumber, soil, plants, seeds, and mentorship to those interested in cultivating their own food.

Initially, our work was focused on self-sustenance. But, over time, we came to realize that being able to feed our families depends on healing the entire food system. We came to understand that the history of the

US food system is rooted in stolen land, exploited labor, and an extractive relationship with the Earth.

The harm began when the white Christian "doctrine of discovery" authorized settlers to perpetuate genocide, enslave millions, and concentrate land and power. That legacy endured, and to this day we live under food and land apartheid, with over 95 percent of the farmland in the United States being white owned.[2] Black farmers currently operate around 1.5 percent of the nation's farms, down from 14 percent in 1910, and have lost over twelve million acres to USDA discrimination, racist violence, and legal trickery.[3] Around 80 percent of the people working the land in the United States are Hispanic, yet only 2.5 percent of farms are owned and operated by the Hispanic community. So, managing a farm is among the whitest jobs in America.[4] Making matters worse, agricultural workers in the United States are not protected by the same labor laws as other workers and do not have the right to overtime pay, days off, fair wages, or collective bargaining.[5]

As a result of all these disparities, hunger, diabetes, heart disease, kidney failure, and other diet-related illnesses unfairly burden people of color. Black and Brown communities remain disproportionately landless and without access to life-giving foods. The food system's oppression of the people is mirrored in its harm to the Earth, with agriculture among the leading causes of greenhouse gas emissions, water shortages, and land degradation.

The more I learned about all these problems, the more I realized how urgently we need to reform our national system of food production and distribution. Soul Fire Farm responded by maturing our work to address the root causes of food injustice, expanding to include farmer training in Spanish and English, policy advocacy, and the incubation of a regional land trust and national Black-Indigenous farmer fellowship. The farm now bustles with two-thousand-plus aspiring farmer trainees each year who go on to take leadership in food and land justice initiatives. We learn from and partner with our elders in the food sovereignty movement, leaders like the Federation of Southern Cooperatives, National Black Food and Justice Alliance, and HEAL Food Alliance.

The work to create a just food system is monumental and daunting. We need to heed calls for specific forms of change, including Indigenous

seed sovereignty, climate justice, and the rematriation of land. Rematriation refers to the return of land to Native nations in a way that centers
the leadership of women and genderqueer people. In times of doubt, I call
to mind my ancestral grandmothers, who hid away their seeds of okra,
cowpea, millet, and black rice in their braids before being forced onto
transatlantic slave ships. Their deep yearning was to have the means to
feed their own children. If they, in those unimaginable circumstances,
had the audacious hope to set aside some seeds for me, who am I to give
up on my own descendants? How could I not plant these seeds for all of
our children?

In the pages that follow, we'll explore five concrete strategies we need
to engage in if we are to plant seeds of justice for generations to come. In
each case, there are specific things all of us can do to help plant and nurture seeds of change for a better food future.

1. FARMWORKER JUSTICE

Reforming our food system must start with changing how we treat the
people whose hands pull our food from the earth. Over three-quarters
of these workers are "foreign-born," which means they are subject to the
myriad oppressions of our broken immigration system.[6]

In 1917, the United States and Mexico inaugurated the first of several
so-called bracero programs, which would bring hundreds of thousands of
manual laborers to the agricultural fields of the United States. The program was later replaced by the H-2 visa in 1952 and the H-2A visa in
1986. The workers who enter the United States under these programs are
seeking a better life for themselves and their families. But farmworkers,
whether domestic or foreign, are excluded from many protections under
the National Labor Relations Act (NLRA) and Fair Labor Standards Act
(FLSA), such as collective bargaining rights, overtime limits, child labor
restrictions, and workers' compensation insurance. Farmworkers often
experience wage theft, unsafe living conditions, workplace harassment,
and lack of access to medical care. Many farmworkers receive wages based
on "piece rate"—for example, eighty-five cents per ninety-pound box of
oranges—rather than an hourly wage, which contributes to one-third of

farmworkers earning less than minimum wage.[7] Helping to lock this network of unfair practices in place is the fact that large corporations now control 50 percent of the food production in this country. They use their power to push to keep farm labor cheap to maximize profits.[8]

Seeds of Change

The people who feed our families deserve full protection under NLRA and FLSA, including a living wage, safe housing and transportation, breaks, overtime pay, workers' comp and unemployment insurance, protection from pesticide exposure, and the right to collectively bargain.[9] Making this happen will require action by Congress, which involves a long-term political struggle. In the meantime, we can encourage farms to uphold worker dignity by buying produce from businesses that are Food Justice Certified or that participate in another domestic fair-trade certification program.[10]

We also need to support programs that seek to empower farmworkers. Many are expert agriculturalists and should be able to advance to management and proprietorship of farms, becoming decision-makers and owners on the land where they labor. Programs like ALBA, Flats Mentor Farm, Groundswell Incubator Farm, and Soul Fire Farm, which focus on training farmers of color, need financial backing to expand their impact.[11] Immigration reform is necessary to create pathways to full citizenship for farmworkers and their families.

2. LAND REDISTRIBUTION

Theft and enclosure of land is part of the DNA of the United States. European colonizers seized 1.5 billion acres of land from Native Americans in a campaign of genocidal violence.[12] African Americans were also victims of land loss through government discrimination, lynching, restrictive covenants, redlining, corporate land grabs, and heirs' property exploitation. Over one-third of Black land is held in heirs' property, where descendants of the original landowner hold the deed in common because of lack of access to legal services. Heirs' property is more vulnerable to court-ordered sales than property with "clean title," and it is one of the leading causes of involuntary Black land loss today.[13]

The long-term impact of these practices has been devastating. In 1920, 14 percent of all landowning US farmers were Black. Today, however, fewer than 2 percent of US farms are controlled by Black people, reflecting a total loss of over fourteen million acres, and over 95 percent of the farmland in this county is white owned.[14] In 1982, the US Commission on Civil Rights acknowledged that discrimination from the USDA was a primary reason Black farmers were dispossessed from our land.[15]

The growing disparities between white and Black people in land ownership in this country mirror and expand the widening racial wealth gap, which has increased to an eight-to-one ratio.[16] Ralph Paige of the Federation of Southern Cooperatives put it simply: "Land is the only real wealth in this country and if we don't own any we'll be out of the picture."

Seeds of Change

As we think about land redistribution, the return of land to Indigenous people must be a top priority. We need to heed the call of the United Nations to share the land back, so that all communities can have the means of production for food security.[17] Recognizing that this is stolen land, rematriation is profoundly and simply giving the land title back, with particular attention to the leadership of women and trans Indigenous people.

The rematriation movement is not just a good idea; it is actually beginning to happen. In 2016, the Sogorea Te' Land Trust rematriated its first property in Huchiun territory, a traditional village site named Lisjan, which is now home to ceremonial grounds, gardens, emergency water catchment, and youth programs.[18]

There's a lot we can do to encourage and strengthen this movement. At a community level, we should support the #landback movement and Shuumi land tax initiatives. Settlers can give their property deeds to Native land trusts or directly to tribal entities, as in the case of the Ponca tribe accepting the return of their land along the Trail of Tears in 2018. At a federal level, expanded funding is needed for the Indian Tribal Land Acquisition Loan Program and the Highly Fractionated Indian Land Loan Program.[19]

Land rematriation is part of a philosophical shift away from the Western idea of private property and toward the Indigenous concept of "land commons." As Julius Nyerere, the Tanzanian anti-colonial activist and political leader, said, "To us in Africa land was always recognized as belonging to the community. Each individual within our society had a right to use the land, because otherwise he could not earn his living and one cannot have the right to life without also having a right to some means of maintaining life. But the African's right to land was simply the right to use it; he had no other right to it, nor did it occur to him to try and claim one."

Land trusts are a powerful tool in the work to return lands to the commons. A land trust is an organization that holds land or conservation easements to protect properties for a public or environmental purpose. The first community land trust, New Communities Inc., was founded in 1969 by Black civil rights activists who were being violently attacked for their voter registration work. They purchased 5,700 acres of land in Georgia, shared among five hundred families, and established residences, cooperative farms, and a community center. Building on this legacy, the Northeast Farmers of Color Land Trust, established in 2017, is working to purchase two thousand acres of land over the next five years to return to Indigenous, Black, and people of color farmers and land stewards.[20] This land will be permanently protected through agricultural and cultural easements.

You can foster this movement by supporting Black and Indigenous land trusts and land commons projects like Black Family Land Trust, Eastern Woodlands Rematriation Collective, Native American Land Conservancy, and White Earth Land Recovery Project, among myriad others.[21]

3. FOOD ACCESS WITH DIGNITY

Systems of oppression ultimately rely on various forms of violence to maintain power and control. An obvious example in the United States is mass incarceration, which disproportionately impacts people of color. Less obvious—but highly insidious and pervasive—is the flooding of our communities with foods that kill us. In its impact, the systematic denial of access to healthy foods is another form of state violence against our people.

As a result, Black people are ten times more likely to die from diet-related illness than from all forms of physical violence combined.[22] The incidence of diabetes, kidney failure, and heart disease is on the rise in all populations, but the greatest increases have occurred among people of color, especially African Americans and Native Americans. From the corner store and the public school lunchroom to the prison cafeteria, our federal government is subsidizing the processed foods that undermine the health and future of our communities. The USDA invests $130 billion annually into industrial agriculture and commodity foods, such as wheat, soy, milk, and dairy, and comparatively little into "specialty crops" like vegetables.[23] Fast-food chains and junk food corporations disproportionately target their advertising to children of color, further fueling the epidemic of childhood diabetes.[24]

Most egregious of all, in this wealthy nation, about thirty-four million Americans are food insecure, and about thirty-nine million people live in US Census tracts that are both low income and lack access to healthy food outlets.[25] This trend is not race neutral. One in five Black and Latino children and one in four Native children go to bed hungry at night.[26] Roughly 17.7 percent of predominantly Black neighborhoods have limited access to supermarkets, compared to 7.6 percent of largely white neighborhoods.[27] This lack of access to life-giving food has dire consequences for our communities.[28]

Clearly, the current food system has not been designed with the best interests of our community in mind. We believe that the term "food desert" is too passive to describe the inequity in today's food system. Our mentor, Black farmer-activist Karen Washington, taught us to recognize America for what it is, a human-designed "food apartheid" system where certain populations live in food opulence while others cannot meet their basic survival needs.

Seeds of Change

Healthy food is a basic human right, not a privilege to be reserved for the wealthy. To honor this right, we need to fully fund the federal government's Supplemental Nutrition Assistance Program (SNAP) and make it easier to use by permitting online purchasing and providing higher allowances. But this is just one part of a complete program to make healthy food available to all. The US health-care system should be updated to allow doctors to prescribe vegetables and fruits, not just pills, and have that "medicine" covered by insurance.[29] Community institutions like schools, hospitals, day cares, prisons, and senior centers need reliable funding to provide whole, real foods to their populations.

We can also treat corporate food-and-beverage companies like tobacco companies: We should hold them liable for adverse health impacts on people through taxation and include visible warning labels. We should end corporate marketing of highly processed food and food brands to children, including in schools, and we should end subsidies for processed food marketing by closing the tax loophole that allows corporate write-offs for this marketing.

Ultimately, we need to work toward food sovereignty, where all people exercise the right to control their own food systems rather than having their access to food controlled by others.

Food sovereignty isn't a novel concept. Black and Brown communities have a rich history of providing food for our families and one another in a dignified manner. Long before and after victory gardens, people of color were growing food through provision and community gardens. In the 1970s, the Black Panther Party fed over twenty thousand children free breakfast nationally, which became a model for school meal programs in the United States.[30] Food hubs, CSAs, and cooperatives all have roots in Black farming communities.

Rather than relying on profit-driven corporations to run the food system, we need to strengthen the food distribution models championed by those most impacted by food insecurity. We can provide capital, credit, tax breaks, and training to worker- and community-owned cooperative food enterprises that generate wealth for our people. It's time to catalyze community support for grassroots farm and food projects that distribute to the most vulnerable populations. Organizations like Sweet Freedom Farm, D-Town Farm, Fresh Future Farm, Rock Steady Farm, and Corbin Hill Food Project have models for getting food directly from the farmer to those who need it most.[31]

4. REPARATIONS FOR BLACK FARMERS

"Imagine your neighbor stole your cow. A few weeks later the neighbor comes over, laden with remorse, to offer a sincere apology and a promise to make it right. The neighbor offers to atone by giving you half a pound of butter every week for the rest of the cow's life. What do you think of that?" The challenge was issued by Ed Whitfield, board member of the Southern Reparations Loan Fund, during the E. F. Schumacher Center lectures in 2018.[32] His audience was unanimous in its response: "We would want our cow back!"

And the United Nations agrees. The UN principles on reparations and immunity, which provide basic guidelines for addressing gross human rights violations, hold that "reparation should be proportional to the gravity of the violations and the harm suffered."[33]

Unfortunately, when it comes to the egregious harm caused by genocidal violence against Native Americans, centuries of American slavery, and generations of Jim Crow, proportional reparations have never been attempted. To adopt Whitfield's parable, American society's scant attempts to make amends for these atrocities is the butter; reparations is the cow.

Real compensation for unpaid wages under slavery alone would add up to $5.9 trillion in today's dollars. That doesn't include damages due Black people because of post-slavery policies such as redlining, mass incarceration, housing and educational discrimination, and other injuries.

Discussions around reparations in this country have been of special interest to Black farmers.[34] The litany of societal abuses heaped upon them includes the broken post-emancipation promise of "forty acres and a mule," lynchings that targeted landowners, discrimination by the federal government, and heirs' property exploitation.[35] As a result, Black farmers have lost more than twelve million acres of land, equaling $120 billion worth of stolen intergenerational wealth.[36]

The victims of this massive crime have been fighting back. In 1997, Black farmers drove their tractors to Washington demanding justice, and they sued the federal government for its leading role in their oppression. The *Pigford v. Glickman* settlement of 1999, widely lauded as the largest civil rights discrimination payout in the country's history, awarded about $2 billion, with a typical disbursement of $50,000 to an individual farmer.[37] While significant, it still fell far short of proportional reparations. The amount did not come close to what is owed and was not enough to buy back the lost acres or pay off the crushing debt farmers had accumulated in their bids for survival.

Seeds of Change

Land and wealth must be redistributed to the descendants of those from whom it was stolen. A first step is to pass H.R. 40, a federal bill that would establish a commission to study reparations proposals for African Americans and recommend appropriate remedies. Meanwhile, since land loss in the Black community is ongoing, we need to intervene by supporting land trusts to purchase and temporarily hold Black land under dispute. Organizations like the Land Loss Prevention Project need our backing to provide legal assistance to Black farmers trying to keep their farms alive.[38] Further, our regressive tax code perpetuates the wealth gap, with rules like "step-up" that allow wealth holders to pass assets on to their children without paying taxes on the full value. A fair tax code could return stolen wealth to people of color through programs like universal basic income or "baby bonds," which would guarantee a minimum ongoing wage and a lump sum wealth transfer to each individual when they reach adulthood.[39]

Federal food programs also need a dramatic overhaul in the interests of racial equity. The Farm Bill covers most policies related to food and agriculture in the United States and currently costs about $156 billion annually. Almost all of the agricultural payouts benefit large corporate farms growing commodity crops, with very few small-scale, sustainable farmers and farmers of color benefiting.[40] The health of the Black farming community will depend on access to capital, credit, training, debt relief, and land—all provisions outlined in the proposed Justice for Black Farmers Act.[41]

To support aspiring farmers of color, we also need to provide full scholarships to land-grant universities and other agricultural degree programs. Currently, many of these programs are geographically inaccessible to people of color; to remedy this problem, satellite campuses on urban and rural farms owned by Black, Latino, and Indigenous people should be developed. These training programs must explicitly address racism in the food system and provide support for healing from land-based trauma. Additionally, passage of the Urban Agriculture Act of 2016 would allow urban farmers to be counted in the Census and receive government support for the work they do to feed the community.[42]

Finally, the USDA 2501 Program (officially the Outreach and Assistance for Socially Disadvantaged Farmers and Ranchers and Veteran Farmers and Ranchers Program) needs to be made more accessible by increasing the funding, decreasing the onerous application and reporting requirements, and offering technical assistance to access the funds.[43]

5. ECOLOGICAL STEWARDSHIP

Finally, any serious effort to reform our food system must address the ecological damage we've created over centuries of mismanagement. As European settlers displaced Indigenous peoples across North America in the 1800s, they exposed vast expanses of land to the plow for the first time. It took only a few decades of intense tillage to drive over fifty varieties of original organic matter from the rich prairie loam soils. The productivity of the US Great Plains decreased by 71 percent during the twenty-eight years following that first European tillage. The initial anthropogenic rise in atmospheric carbon dioxide levels was due to that breakdown of soil organic matter. Land clearing and cultivation emitted more greenhouse gases than the burning of fossil fuels until the late 1950s.[44]

Western industrial agriculture remains a leader in anthropogenic environmental impacts, accounting for 26 percent of greenhouse gas emissions, 50 percent of land use, 70 percent of freshwater withdrawals, and 78 percent of global ocean and freshwater eutrophication.[45] At the same time, US consumers spend less on food, as a percentage of income, than any other country in the world.[46] These two phenomena are closely linked. Our expectation of cheap food results in the real costs of food production being externalized and passed on to the Earth and vulnerable people in the form of climate change, damage to soil and water, health-care costs, and low wages. This needs to change.

Seeds of Change
Along with seeds, our grandmothers braided ecosystemic and cultural knowledge of land care into their hair before the forced transatlantic journey. They braided the wisdom of sharing the land, as captured in practices like the *huza* farm co-op system of the Krobo people. They braided the

wisdom of sharing labor and wealth, as in the *dokpwe* worker co-ops and the *susu* credit unions of the Dahomey people. They braided the wisdom of caring for sacred earth, as in the carbon-capturing dark earth compost of Ghana, the non-till raised beds of the Ovambo people, the soil-stabilizing polycultures of Nigeria, the rotational grazing and swidden agriculture of Guinea, and the *zai* and *tassa* pits that turned the Sahel desert green.

This legacy of sustainable agriculture was carried on in the African diaspora. Dr. George Washington Carver, Black agronomist at Tuskegee University, believed, "Unkindness to anything means an injustice done to that thing. . . . The above principles apply with equal force to soil." He spent his predawn hours in the forest listening to the divine voice of the plants to give him instructions for his research. Arguably one of the founders of the modern organic and regenerative farming movement, Carver advanced leguminous cover cropping, mulching, composting, and crop rotation and restored severely degraded southern soils. Carver—along with Booker T. Whatley, Fannie Lou Hamer, Shirley Sherrod, and others—established a foundation for culturally rooted regenerative agriculture.[47]

Despite immense land-access challenges, Black farmers today persist in using heritage agrarian practices to capture carbon from the air and trap it in the soil. Their strategies include silvopasture and regenerative annual cropping, solutions ranked number nine and number eleven, respectively, in Project Drawdown.[48] Project Drawdown quantified and analyzed over one hundred climate solutions for their potential impact on the climate crisis. Silvopasture is an Indigenous system that integrates nut and fruit trees, forage, and grasses to feed grazing livestock. Regenerative agriculture involves minimal soil disturbance, the use of cover crops, and crop rotation to increase soil organic matter. Both systems harness plants to capture greenhouse gases, a mechanism that ecologist Paul Hawken calls the most effective tool we have for addressing global warming. For every 1 percent increase in soil organic matter, we sequester roughly 8.5 metric tons of atmospheric carbon per acre. So, if we were to scale these ancestral farming practices globally, we could pull over one hundred gigatons of carbon dioxide out of the atmosphere and put it back into the soil where it belongs.[49]

To make regenerative agriculture economically viable, we need to massively strengthen the federal Environmental Quality Incentives Program (EQIP) to pay farmers for restoring ecosystem services and stewarding the public trust of soil, water, and atmosphere.[50] Preferably, the scaling of EQIP will be funded by taxing corporate farms that are driving climate change, extinction, and soil erosion. Directly supporting Indigenous and Black regenerative farms is also a crucial step in advancing ecological land stewardship.[51]

HOPE IS A SEED

Since its inception, this nation has relied upon the labor, expertise, and resources of BIPOC (Black, Indigenous, and people of color) communities to undergird the food system. Even today, without the labor of people of color, food production would screech to a halt. We people of color are essential, not just for our labor but for our lives. We hope that this becomes a moment of awakening to the truth that "to free ourselves, we must feed ourselves." All of us deserve this freedom.

As we work toward a racially just food system, it's crucial that we don't adopt the same colonizer mentality that created the problems in the first place. The communities at the front lines of food justice are composed of Black, Latino, and Indigenous people, refugees and immigrants, and people criminalized by the penal system. We need to listen before we speak and follow the lead of those directly impacted by the issues.

People of color invented many of the solutions championed by the good food movement today. Booker T. Whatley brought us CSAs. Dr. George Washington Carver codified and spread regenerative (organic) agriculture. Shirley and Charles Sherrod started the first community land trust in the United States. Organizations like the National Black Food and Justice Alliance continue to develop novel solutions and need our support.[52]

It is not enough just to emulate Afro-Indigenous agricultural practices or those from Indigenous communities across the globe. We must have the courage to address generational trauma from centuries of land-based oppression. While the land was the scene of the crimes, the land was never the criminal. By working together on land reform, farmworker rights, reparative justice for Black farmers, and rematriation of land for Indigenous communities, we can ensure that those with essential ecosystemic and cultural "re-membering" have the resources to implement the land-honoring practices that we desperately need.

Discouragement is not an option in these times. Our ancestors never gave up on us, and we must not give up on our children. Let us also braid seeds of hope in our hair as we journey across the tumultuous waters of our own times.

6.

The Environmental and Community Impacts of Industrial Animal Agriculture

By Lisa Elaine Held

Lisa Elaine Held is the senior staff reporter at *Civil Eats*, a daily news source for critical thought about the American food system. Since 2015, she has reported on agriculture, the food system, and food policy with a focus on climate, environmental justice, inequality, and health. Her stories have appeared in publications including the *Guardian*, the *Washington Post*, and *Mother Jones*. She also hosted and produced more than one hundred episodes of the podcast *The Farm Report* on Heritage Radio Network from 2018 to 2021. Previously, she covered health, wellness, and nutrition for publications including *Well+Good* and the *New York Times*. She has a master's degree from Columbia University's School of Journalism and is based in Baltimore, Maryland.

Gina Burton is part of a very big family rooted in a very small state. Her grandmother had eighteen children; her grandmother's sister had thirteen. When Gina was a child in Millsboro, Delaware, in the 1970s, the family reunions were so large, they used all five of the adjoining properties on Herbert Lane to gather together.

She still lives in one of those houses, which form a row of simple, single-story ranches on a narrow street. Her mother, aunt, and cousins live in the others. Her sister recently built a bigger home at the end of the lane, in the spot where her grandmother's house once stood.

Still, Gina's family members have always been vastly outnumbered by chickens.

In 2017, poultry companies in Sussex County—which covers about one thousand square miles, including Millsboro—produced 193 million chickens for meat.[1] In what is essentially the Burton family's front yard, Mountaire Farms, the fourth-largest chicken producer in the United States, slaughters and processes two million of those chickens per day.[2] Then the company sprays the wastewater from that process onto farm fields that stretch right up to the road in front of the Burton family homes. On one November day in 2022, the irrigation equipment was sitting so close to the street that it would have sprayed a passing vehicle had it been operating.

Gina's grandfather worked for Townsends, Inc., the company that operated the Millsboro plant before Mountaire took over in 2000, so the Burton family story has been tied up with the industry for generations. But over the last decade, Americans' ravenous appetite for crispy chicken sandwiches has altered their lives in more profound, destructive ways.

By November 2022, because of a court settlement that included a gag order, Gina couldn't talk to me about Mountaire specifically. But the legal complaint that she and one hundred of her neighbors filed in 2018 had already laid out their allegations in alarming detail, and many of the allegations are supported by federal and state agency records and significant news coverage.

According to the complaint, for more than a decade, the wastewater Mountaire had been spraying on the surrounding fields contained dangerous levels of nitrogen and bacteria.[3] Nitrates—a chemical form of nitrogen—seeped into the groundwater and ended up in surrounding wells at levels up to nine times those considered safe for drinking water, according to a violation notice from the Delaware Department of Natural Resources and Environmental Control (DNREC).[4] DNREC's notice showed that the wastewater also contained fecal coliform bacteria at levels that can cause serious illness.[5]

Initially, the Burtons had no idea this was happening. In 2017, representatives from Mountaire began leaving bottled water on the Burtons' and other neighbors' doorsteps, but they didn't communicate any of the risks of using the well water. That's when the family started to get suspicious.

High levels of nitrates in drinking water can cause birth defects, thyroid disease, and a deadly infant disorder called blue baby syndrome.[6] They also increase the risk of certain cancers.[7] Fecal coliform bacteria vary in their impacts, but many, including *E. coli*, can cause upset stomachs, vomiting, diarrhea, and fever.[8]

"When you looked at the record, those fields were saturated with nitrates for years," said Chris Nidel, a Washington, DC–based attorney who helped the Millsboro residents sue Mountaire. "[This kind of abuse] happens in the chemical [industry] context often, but people don't think of agriculture in the same way, as far as causing acute and chronic environmental illness and injury."

In fact, a handful of large industrial animal-agriculture companies producing chicken, pork, and beef all over the country regularly cause air and water pollution that harms people, wildlife, and critical ecosystems. And compared to other industries, agriculture is not as closely regulated under environmental laws including the Clean Water Act and the Clean Air Act. In Nidel's view, "It's just anything-fucking-goes."

To be clear, as a country, we do not have to produce chicken or any other meat this way. Over the past several decades, companies in search of bigger profits seized opportunities to apply the efficiencies and scale of industrialization to meat.[9] They found ways to make animals grow faster and bigger, crowded as many as possible indoors in concentrated animal feeding operations (CAFOs), and built assembly-line-style processing plants.[10] Farmers were faced with the choice of falling in line, struggling around the margins in local food systems, or getting out of agriculture altogether. To wit, in the 1970s, the US secretary of agriculture literally told them to "get big or get out."[11]

Scale is not always bad. In fact, a certain minimum size is necessary for farm businesses to survive and thrive and to ensure a secure food supply. But there is a point at which the balance shifts. Over the past few

decades, a shrinking number of bigger and bigger companies have consolidated their power and maximized efficiencies to get as big as possible and
produce cheaper and cheaper meat.[12] Along the way, they have wielded
money and influence to create a narrative: cheap meat is necessary to feed
the world.

While that fact is debatable from many angles, the data is clear on one
point: Americans have happily gobbled up what the industry is selling.
Over the past several decades, as prices dropped and attention turned to
the health (and, later, climate) impacts of eating too much beef, consumption of red meat fell while consumption of chicken increased. Americans
are projected to eat more than one hundred pounds of chicken each in
2023, almost quadruple what they ate in the 1960s.[13] Costco's $4.99 ro-
tisserie birds have a cult following, and food writers breathlessly chronicle
which fast-food chain's crispy chicken sandwich is king.[14]

Chickens that grew to 3.4 pounds over sixty-three days in 1960 now
bulk up to 6.5 pounds in just forty-seven days on farms that can house up
to a million—instead of a few thousand—birds at a time.[15] But animals
are not widgets, and there are consequences to treating them as if they
were. Left out of the story are the costs shifted onto the ecosystem and its
inhabitants, those same gracious consumers of ninety-nine-cent nuggets,
as those companies extract as much profit as possible.

In July 2021, the Rockefeller Foundation released a report estimating
the true cost of food in the United States. It found that while Americans
reached into their wallets and paid just over $1 trillion to purchase food
in 2019, the actual cost—including the economic toll of pollution, greenhouse gas emissions, health care, and other factors—was three times that
amount, at more than $3 trillion.[16]

The story of how chicken production has impacted the residents and
environment in Millsboro, Delaware, is just one example of the costs associated with our industrialized meat production system across the country.

RUINED RIVERS AND VAST DEAD ZONES

Herbert Lane is a dead end. If the road kept going, it would lead directly into the Indian River, which flows along the southern edge of the

Millsboro fields where Mountaire sprays its chicken-processing waste. Directly across the river on the opposite bank, Wayne Morris has lived in a working-class neighborhood called Possum Point since 1968. On a sunny November day, he's wearing a blue work coat with a patch that says "Millsboro Fire Department." The tip of a wispy, snow-white goatee tickles the center of his chest.

When Morris was young, he trapped soft-shell crabs and caught small flounder in the river's deep water. Now, docks where boats were once tied are six to ten feet above shallow, muddy water that's barely moving. "The river is no more," he says. "It's a joke."

Many factors, including overdevelopment and old, leaky septic systems, have likely contributed to the waterway's decline, but the chicken industry has played an important role.[17] The river snakes past the fields where Mountaire had been over-applying wastewater, right up to another chicken-processing plant owned by Allen Harim. Plant operators there had been discharging wastewater into the river at levels that violated their permits for years, according to DNREC.[18]

According to the court complaint filed by Chris Nidel, the median nitrate concentration of groundwater in the Indian River Bay watershed is about six times what would be considered healthy for a river.[19] High nitrate levels affect oxygen levels, reducing the water's ability to support aquatic life. Near Possum Point, 2020 data found the water failed to meet the state's standard for oxygen on 75 percent of mornings, and volunteers have documented fish kills and dead crabs over the years.[20]

The production of too much nitrogen, as well as phosphorus, is a problem that is built into the CAFO system from the get-go. As local activist Maria Payan says, "It's always the same thing. We can't get rid of the friggin' waste." Hundreds of thousands of animals confined in one place produce an incredible amount of urine and manure that needs to be disposed of. Then, those animals are further concentrated at processing plants, which produce more waste. To get rid of it all, companies dump it on fields at levels that plants can't absorb, and the nutrients end up in groundwater, then in waterways.[21]

Case in point: Mountaire was authorized by DNREC to apply no more than 320 pounds of nitrogen per acre to its fields annually. According to DNREC records, by September 2017, it had already exceeded that level in eleven out of thirteen Millsboro fields. In one field, it applied 1,038 pounds, more than three times the legal limit. Mountaire's wastewater also repeatedly exceeded safe levels for fecal coliform bacteria.[22]

Chicken waste is a big reason why a 2022 Environmental Integrity Project report found that Delaware beats every other state in the percentage of its rivers and streams that are so polluted they are unsafe for human contact and/or unable to support aquatic life.[23] Ninety-seven percent of the state's 1,104 miles of assessed waterways are listed as impaired.

Those rivers and streams flow into larger bodies of water like the Inland Bays and the Chesapeake Bay. The Chesapeake is particularly affected, since the chicken industry hasn't just taken over Delaware.[24] The concentration extends down the Delmarva Peninsula, which stands between the bay and the Atlantic, into sections that belong to both Maryland and Delaware. Salisbury, Maryland, just south of the Delaware border, is the birthplace of Perdue Foods, and Tyson Foods also has a huge presence in the area.[25] In 2021, 567 million chickens were produced on the peninsula.[26] Each year, federal agencies spend hundreds of millions of dollars on efforts to clean up pollution in the Chesapeake Bay, a significant percentage of which comes from agriculture, including poultry.[27]

The only state that has a bigger animal poop problem than Delaware is Iowa. That's because instead of chickens, Iowa is home to twenty-three million pigs at any given moment, an increase of 64 percent in the last twenty years. And pig waste is a real shitstorm compared to chicken waste. In 2019, a University of Iowa research engineer calculated that, while Iowa has a population of just 3.2 million people, it produces the waste of 168 million.[28] In November 2019, the Iowa Department of Natural Resources found that more than half of the state's rivers, streams, lakes, and wetlands failed to meet water quality standards due to pollutants including nitrogen, *E. coli*, and cyanobacteria. The agency has also documented hundreds of manure spills from confinement operations, which regularly kill thousands of fish and other aquatic organisms.[29]

In Minnesota, the number of CAFOs has tripled since 1991, mainly for pork production. In one 2020 report, researchers at the Environmental Working Group (EWG) modeled the amount of waste that would be produced at each CAFO and compared that to what could safely be applied on nearby fields. They found thirteen counties where nitrogen from manure along with chemical fertilizers likely exceeded the recommendations of the Minnesota Pollution Control Agency by more than half. One county where hog production is most concentrated had an estimated 14,368 tons of nitrogen overload.[30]

Smithfield Foods, the world's largest pork producer, owns eleven of the biggest hog CAFOs in Missouri, each of which is permitted to house up to seventy thousand pigs in one place. Between 2012 and 2022, seven of the eleven have received warnings or violation notices from the state's environmental agency. In March 2022, one of the facilities spilled approximately 350,000 gallons of hog manure and wastewater, which polluted between twelve and fifteen miles of local waterways. In a press statement, a Smithfield spokesperson said the spill was an "extremely rare incident" and that the company had modified its equipment and systems so that it wouldn't happen again, adding that the company "takes pride in its longstanding compliance record" in Missouri.[31]

Excess nutrients from Iowa, Minnesota, and Missouri also have far-flung impacts, since some keep moving from groundwater and local waterways into the Mississippi River. When they get to the Gulf of Mexico, they contribute to the annual dead zone, an area where aquatic life cannot survive at all. Between 2018 and 2022, the average size of the dead zone was about 4,300 square miles, which is more than twice the size of the entire state of Delaware.[32] According to the same 2020 EWG report on Minnesota, between 60 and 80 percent of the nitrogen in the dead zone can be traced to manure and synthetic fertilizer runoff from Midwest farms.

DON'T DRINK THE WATER.
KEEP YOUR WINDOWS SHUT.

In Millsboro, residents miss having the Indian River as a source of recreation and natural beauty. But their biggest concern over the past decade has been whether they can safely drink, cook with, or bathe in their own water.

Most homes in the town, like Gina Burton's, have private wells. According to the complaint filed against Mountaire, in 2017, groundwater monitoring wells near Mountaire's fields consistently tested above the drinking-water standard of ten milligrams per liter of nitrates. In some tests, levels reached as high as 92.5 milligrams per liter, more than nine times the level considered safe. Levels of coliform bacteria were also consistently unsafe.[33] Over several years, in response to DNREC enforcement actions, this lawsuit, and another related lawsuit, Mountaire officials have said that the company's operations are not the cause of elevated nitrate levels in private wells and have pointed to statistics that suggest the county has historically had elevated nitrates in its soil.[34] In a 2021 statement, a spokesperson said, "Over 100 years, our company has made the communities where we live and work a top priority, and we take our neighbors' concerns very seriously."[35]

Residents, including the Burtons and their neighbors who joined them in filing the lawsuit against Mountaire, told reporters at the time that they had been drinking their well water for years before learning of possible contamination. As a result, the legal complaint is filled with a long list of medical ailments they attributed to that exposure. Many suffered from gastrointestinal distress; one man had a portion of his intestine removed. Multiple women reported miscarriages; one woman's children suffered birth defects and cognitive impairment.[36] Of course, it is impossible to prove a direct causal link between each person's exposure to the contaminated water and the various illnesses they reported, but there is well-established science that shows the level of exposure they experienced increases the risk of many of these conditions.

In 2020, Mountaire chose to settle with the families for an undisclosed amount.[37] But, as of December 2022, Gina and others were still waiting for a permanent solution to their water woes. Mountaire employees were still showing up every two weeks to deliver filtered water tanks for each home. "It's been three years," Gina said. "Nobody I know has got deeper wells or new wells or got on the public water system. That hasn't been solved."

In Iowa, many of the state's wells have also been found to be contaminated with both nitrates and coliform bacteria (like *E. coli*). In one pork-heavy county in the northwest part of the state, 115 of 779 wells tested had elevated nitrate levels, and 411 tested positive for bacteria between 2002 and 2017. In early 2020, many residents there had switched to drinking only bottled water just to be safe.[38] In the capital city, the Des Moines River is residents' main source of drinking water. The city of Des Moines has been struggling to deal with nitrate pollution, much of which comes from hog CAFOs, in the river for more than twenty years. After a few dry years offered a brief reprieve, the city was forced to turn on a nitrate removal plant in 2022, which can cost local taxpayers up to $10,000 per day to run.[39]

Gina's son Kiwanis was a likely victim of another form of pollution caused by industrialized food production. According to his obituary, Kiwanis Burton was "a shy young man with a beautiful smile. He was known for his basketball skills and his love for eating."[40] Kiwanis also had severe asthma that he struggled to keep under control.

Gina knew how bad it was. She dealt with asthma herself, and she described the worst attacks as "like somebody put a plastic bag over your head and was just twisting the bag to the point where you can't breathe." She watched as the number of shots Kiwanis needed each week increased.

Still, when she got the call from the hospital saying his condition was critical, "I still couldn't grasp the fact that somebody was dying of asthma," she said. Especially a twenty-four-year-old.

In addition to contaminating water, industrial animal agriculture creates pollutants that are released into the air. Those pollutants also mainly come from the concentration of waste, which when accumulated can emit ammonia and nitrous oxide, particulate matter, hydrogen sulfide, and volatile organic compounds.[41] In chicken and pork CAFOs, farmers use massive fans to blow those pollutants out of the barns so the animals can breathe. Humans nearby bear the brunt. Pollutants also get into the air when the waste slurry is spread on fields, like those that stretch right up to the Burtons' front doors.

Inhaling those pollutants is dangerous at certain levels, and some research shows that for people living near CAFOs, the risk of respiratory conditions like asthma, bronchitis, and pneumonia rises. One study published in 2021 estimated that thirteen thousand deaths each year can be directly attributed to emissions from industrial animal agriculture.[42] A 2017 study found residential proximity to CAFOs was associated with asthma medication orders and hospitalizations, while a 2015 review of the available scientific literature found consistent correlations between living near CAFOs and respiratory problems, including asthma and chronic obstructive pulmonary disease, as well as infections with MRSA, high blood pressure, and other health problems.[43]

Most of the evidence so far comes from research done on communities in Duplin and Sampson Counties in North Carolina, where Smithfield's hog farms have polluted waterways and ruined the health and quality of life in Black communities for decades.[44] But one study published in 2018 looked specifically at poultry CAFOs in Pennsylvania and found that people with high rates of exposure to emissions from the farms were 66 percent more likely to be diagnosed with a form of pneumonia.[45]

Kiwanis was one of several Millsboro residents in the complaint who reported suffering from asthma and other respiratory illnesses. Just south, in Wicomico County, which has the most concentrated chicken production in Maryland, the childhood asthma rate between 2005 and 2009 was 25 percent, compared to a national rate of 8 percent. Respiratory disease is 54 percent higher than the state average. Residents there have been fighting to get air-monitoring systems put in place for years, since many suffer from respiratory ailments but haven't been able to prove a direct link between the air they breathe and the chicken CAFOs that surround them. Poultry companies deny responsibility. For example, they often point to higher rates of smoking in the county compared to average as a possible cause of the high rates of respiratory disease.[46]

The question of causation is complicated, but any visitor passing through the area can attest to air quality problems. Drive by a cluster of metal chicken barns on a summer day with your windows down, and an irrepressible, hacking cough will overtake you. It can feel like when a red pepper flake gets caught in the back of your throat, except instead of a speck, the irritant coats the entire surface area and hangs on tight. Those who live with it don't need further proof. "You don't enjoy the air like you did when you was younger," Gina said. "These days, you've got to keep your windows shut."

INDUSTRIALIZED MEAT AND THE CLIMATE CRISIS

Water and air pollution from CAFOs mainly affects isolated groups of low-income people in small towns cut off from most of the American population. But no one is immune to the impacts of the climate crisis, which is also being accelerated by industrial animal agriculture.[47]

In the series of reports issued in 2022 by the world's most trusted scientific panel of climate change experts, those experts directed new attention to the food system. Industrial animal agriculture is at the core of that system's climate flaws.[48]

It starts with feed. Producing food for billions of animals requires enormous resources. Many pesticides used in feed production are fossil-fuel based, and chemical fertilizer used to grow crops releases nitrous oxide—a greenhouse gas three hundred times more powerful than carbon dioxide—when spread on fields.[49]

In the CAFOs themselves, there is the manure that, when spread on fields, can also produce nitrous oxide and methane. Pork and dairy CAFOs are particularly problematic because most store pig and cow manure in lagoons. That concentrated waste emits methane, which is about thirty times more potent than carbon dioxide when it comes to warming the planet. While meat production has gotten more efficient in some ways over time, methane emissions from manure lagoons have risen significantly over the past thirty years, according to Environmental Protection Agency data.[50]

Still, industrial beef and dairy have the biggest climate impacts because of land use and a process called enteric fermentation. It's a fancy word for cow burps, which also emit methane.[51] Of course, cows belch methane no matter how you raise them. And some proponents of industrial systems argue that raising them quickly in feedlots (cattle CAFOs are generally large, crowded outdoor pens, not metal barns) means they burp less methane compared to letting the cattle live longer on grazing land.[52] Some say that soon we'll be able to put additives in cattle feed that reduce methane burps.[53] And others claim that if cattle are raised outside in rotationally grazed pastures, methane can be offset by the healthy soil built during grazing, which can put carbon back into the ground.[54] It's a debate filled with nuances based on individual approaches, geography, climate, and soil type, and the research isn't settled on the exact greenhouse gas comparisons.

Yet we know that no matter how cattle are raised, beef is a more climate-intensive food than grains or vegetables.[55] Advocates for veganism say that's all that matters. Advocates for meat eating will say that the comparison isn't fair, because meat is so much more nutrient rich, and that we need grazing cattle to build healthy soil. Some food-and-climate advocates even argue that raising chickens in CAFOs has helped slow the climate crisis, since chicken is much less climate intensive than beef and having so much cheap chicken shifted Americans' appetites in that direction.[56]

All of those arguments may be worthwhile, but the bigger picture remains. Industrial animal agriculture has increased overall meat eating in rich countries like the United States to unsustainable levels, causing increased greenhouse gas emissions, water and air pollution, and widespread harm to people, communities, and critical ecosystems. A sustainable future requires less meat produced this way, whether it's beef, pork, or chicken. Gina Burton shouldn't have to suffer day after day because boneless, skinless breasts produce fewer planet-warming emissions compared to a burger.

A BETTER WAY?

Across the Chesapeake Bay, way out in the rolling hills of western Maryland, Ron Holter has been working on the antithesis of CAFO production for twenty-five years. Holter moves a small, manageable herd of healthy dairy cows regularly through sixty-six individual paddocks. As the animals graze, they leave their manure behind. Because there are not too many and they're so quickly on the move, that manure doesn't accumulate. So, unlike the managers of chicken CAFOs in Delaware or hog or dairy CAFOs in Iowa, Holter doesn't need to spread the waste on additional fields at levels the landscape can't handle. Instead, it gets incorporated back into the earth, contributing to healthy soil and the regrowth of the grasses in each pasture.

Over the decades he's spent fine-tuning his system, Holter has watched as diversity increased on and below the ground and in the surrounding environment. More varieties of grasses and other plants appeared, the community of microbes in the soil grew, and multiple bird species began nesting in the trees and bushes that frame the pastures. Soil and water stopped running off the fields, so nutrients like nitrogen stay put; they don't end up in neighbors' wells or in the Chesapeake.

And while many will dismiss small-scale production of this kind as a fantasy system that could never be used to feed the world, Holter is not selling his milk at some local roadside farm stand. He's an Organic Valley farmer. His milk is processed with milk from other surrounding small dairies and then sold at the country's biggest grocery chains. He is in fact making a significant contribution to feeding the country, and he's doing it while making thoughtful choices about how to produce food without causing harm.

There are countless other examples of farmers bucking the trend, from grass-fed beef and pastured pork producers in the Northeast to pastured chicken growers in California. There would be more if policies and markets supported them, and there is plenty of room to grow the field. Individually, these farms will always produce less meat and dairy—and that's a good thing.

Gina's life might have been different if her home was facing Holter's farm instead of fields owned by a chicken company intent on churning out as many dirt-cheap party wings as possible. But despite her ongoing struggles, she isn't planning to move away from Herbert Lane, where generations of her family have lived and died and worked and played in cornfields. Recently, she found a new church, and the members of that church have become an even larger family she considers as close as her blood relatives. She has found God in new ways, which has helped her process and handle losing Kiwanis. "God gives you peace and understanding...and I've learned to grasp it better and understand it better. Before I found the new church, I was thinking about moving somewhere else. Now, I'm here and praying that God will keep us safe from...everything," she said, "including pollutants."

7.

A Historic Revolution in the Food Service Industry

By Saru Jayaraman

Saru Jayaraman is an academic at University of California, Berkeley, and the president of One Fair Wage, a national organization working to raise wages for service workers. She was named one of the "CNN 10 Visionary Women," a White House "Champion of Change," a James Beard Foundation Leadership Award winner, and the *San Francisco Chronicle* "Visionary of the Year." Jayaraman authored *Behind the Kitchen Door, Forked: A New Standard for American Dining,* and *Bite Back: People Taking on Corporate Food and Winning* and has appeared on MSNBC, HBO, PBS, CBS, and CNN. She attended the Golden Globes with Amy Poehler as part of #MeToo action in 2018.

As depicted in the film *Food, Inc. 2*, as well as in the other chapters of the book, the system by which food is produced, processed, prepared, distributed, and used in the United States is a deeply dysfunctional one, with serious implications for public health, the environment, and society as a whole. A major aspect of this dysfunction is the treatment of working men and women in many segments of the food system, which sadly is characterized by injustice, bias, and exploitation in many forms.

In this chapter, we'll look closely at the story of labor in just one representative part of the food system: the restaurant business. With the largest
workforce within the food sector, the restaurant industry has long been
a "low road" leader, paving the way to profitability by setting the lowest
possible wage floor for the food system and the economy overall for generations. As you'll see, this tragedy represents one of the many legacies of
slavery that persist in our economy even today, more than a century and a
half after the formal abolition of slavery in the United States.

But today there is a glimmer of hope. In the wake of the COVID-19
pandemic that began in 2020, a historic revolution is turning the restaurant industry on its head, allowing restaurant workers to lead a revolt of
low-wage workers throughout the food system and the country, creating
the possibility that the restaurant industry may become a "high road"
leader for all.

In this historic moment, restaurant workers and their representatives
are organizing and winning unprecedented victories to raise wages and
overcome those persistent legacies of slavery. These victories could fundamentally change long-held assumptions about the nature of work in the
food system.

TIPPING AND THE LEGACY OF SLAVERY IN
THE RESTAURANT INDUSTRY

With nearly fourteen million workers (prior to the 2020 outbreak of the
COVID-19 pandemic), the restaurant industry has been one of the largest and fastest-growing sectors of the US economy.[1] But it has also been
among the lowest-paying industries for generations. One reason is the
practice of tipping, which has become a replacement for wages—a unique
system that exists nowhere else in the world and that makes it extremely
difficult for restaurant employees to earn a living wage.

Tipping originated in feudal Europe, when aristocrats and nobles offered tips as bonuses or extras on top of a salary to serfs and vassals who
worked for them in their homes or in the fields. The concept of tipping
came to the United States in the 1850s, when rich Americans began to
travel to Europe by ship. When they returned home, these Americans began tipping service workers, attempting to show off that they knew the

rules of the European aristocracy and were affluent enough to emulate them.

At first, tipping was resoundingly rejected in the United States. Anti-tipping advocates called it a vestige of feudalism, arguing that, in a democracy, customers should receive good service regardless of how much they could afford to tip and that employers, not customers, should be responsible for paying workers fairly. This rejection resulted in seven states passing complete prohibitions on tipping.[2]

The populist anti-tipping movement spread to Europe, where labor unions picked it up and demanded full, livable wages in lieu of tips, declaring that service workers should not be dependent on the whims and biases of customers. Over the last century, this movement has resulted in European hospitality workers being considered professionals who can attend prestigious schools like Le Cordon Bleu to hone their craft, and who can build a career in hospitality that allows them to feed their families and thrive.

However, in the United States, despite this populist movement, anti-tipping sentiment was largely quelled in the 1850s and 1860s due to restaurant employers' quest to seek ever-cheaper labor. In 1853, waiters in large northeastern cities, who were mostly white men who received full wages and no tips, went on strike to demand higher salaries from their employers.[3] In response, restaurants replaced these male waiters with women entering the workforce, offering them far lesser wages. Twelve years later, when emancipation finally resulted in the slaves being freed, restaurant owners sought an even cheaper source of labor. They hired newly freed African Americans arriving in northern cities after emancipation, offering them service jobs with no wages at all—only the opportunity to obtain white customers' tips.[4]

The restaurant industry was not alone in this practice. The Pullman train company hired tens of thousands of African American men as porters on luxury trains crossing the country and offered them no wage, only the opportunity to obtain white customers' tips. Fighting back against the spread of this legacy of slavery to another sector of the economy, a porter named A. Phillip Randolph organized the Brotherhood of Sleeping Car Porters, the first Black union in the United States.[5] It took many years of strikes and actions, but eventually Brotherhood members won the right to an actual wage rather than having to rely on tips alone.

Meanwhile, the restaurant industry hired mostly Black women as waitresses, similarly offering them no wage and requiring them to live on tips alone. Without a union, these tipped workers were entirely dependent on customers for their income. In 1919, restaurant owners formed the National Restaurant Association, which those of us in the labor movement refer to as "the Other NRA." The organization had the express mission and intent of suppressing food workers' wages in order to keep their costs as low as possible. Twenty years later, in 1938—when the labor laws included in Franklin Delano Roosevelt's New Deal were being passed—the Other NRA successfully lobbied to exclude both farmworkers and tipped restaurant workers from the nation's first federal minimum wage law.[6] Farmworkers were guaranteed no overtime protections when they worked more than forty hours in a week, and tipped restaurant workers were guaranteed no wage at all, based on the idea that they would receive their income wholly in tips. Thanks to this history, the restaurant industry has played an outsize role in setting the low-wage floor for the food system and the entire US economy. Restaurant jobs have consistently ranked as sixth or seventh in the ten lowest-paying jobs in the United States.[7]

In the decades since that first federal minimum wage law was passed, real wages have stagnated for all workers across the US economy, leaving the federal minimum wage at a woefully inadequate level.[8] But in the case of the restaurant industry, matters are even worse. Over time, wages there actually decreased from a full wage prior to 1853 to no wage at all, as women and Black people—and Black women in particular—entered the industry. Since 1938, when the notion that this workforce could be paid no wage was codified into law, the Other NRA has successfully lobbied to suppress these workers' pay. Thus, the restaurant industry's wage structure has been the result of an imbalance of power between one of the nation's strongest employer trade lobbies and a workforce of women of color and single mothers, some of the most vulnerable workers in America.

THE COVID-19 PANDEMIC: A MOMENT OF HISTORIC CHANGE

The COVID-19 pandemic has begun to upend this historic imbalance of power between restaurant workers and employers, with workers moving to reject the subminimum wage for tipped workers for the first time since emancipation. The organization that I lead, One Fair Wage, is dedicated to supporting this long-overdue change.

The subminimum wage for tipped workers is still $2.13 an hour at the federal level. If the overall minimum wage had continued to rise with productivity (as it was required by law to do until 1968) it would be $24 today.[9] Thus, a subminimum wage stuck for decades at the same level is not merely staying the same. In real terms, it is steadily declining, making those who rely upon it poorer and poorer.

Not all states allow employers to impose the federal subminimum wage on tipped workers. But as of 2022, forty-three states persist with this legacy of slavery, allowing employers to pay tipped workers a lesser wage simply because they receive tips. Gender bias is deeply embedded in this difference. Tipped workers continue to be more than two-thirds women, largely working in very casual restaurants and bars. They struggle with three times the poverty rate of other workers and use food stamps at double the rate.[10] They also suffer in other ways. Professor Catherine MacKinnon, the leading legal scholar on the issue of sexual harassment, has published research showing that tipped workers have the highest rates of sexual harassment of employees in any industry because their income is almost entirely dependent on feelings of customers, creating a power dynamic with male customers that makes these workers incredibly vulnerable to harassment in order to feed their families.[11] In surveys, tipped restaurant workers also report other forms of mistreatment—for example, being encouraged by managers to dress in sexually provocative styles in order to earn more money in tips, which makes them even more vulnerable to harassment by customers, coworkers, and managers.[12]

Data also shows that the subminimum wage for tipped workers has resulted in race and gender pay gaps. Tipped restaurant workers who are women of color earn at least $5 an hour less than their white male counterparts due to the segregation of women of color into lower-tipping establishments and implicit bias in customers' tipping behavior toward workers of color, especially women of color. Once again, the fact that these workers are reliant on tips for a portion of their base wage makes them vulnerable to the biases of customers. Research shows that tipping is more correlated with the race and gender of the server than it is with the quality of service.[13]

These practices are not universal. Seven states have rejected the concept of the subminimum wage for tipped workers. Alaska, California, Minnesota, Montana, Nevada, Oregon, and Washington have all required employers of tipped workers to pay a full minimum wage with tips on top since their first passage of a state minimum wage. The economic results from providing one fair wage have been significant. These seven states have the same or higher restaurant sales per capita, restaurant-industry job growth rates, small business restaurant growth rates, and tipping averages as the forty-three states with a subminimum wage for tipped workers.[14] Furthermore, workers in these seven states report one-half the rate of sexual harassment compared to states that limit tipped workers to the federal minimum wage of $2.13 an hour. Evidence also shows that reducing workers' dependence on tips lowers racial income inequities.[15]

Still, most of the nation's food service workers continue to suffer unfair and discriminatory labor and pay practices. The COVID-19 pandemic exacerbated these problems. With the national shutdown in March 2020, six million restaurant workers immediately lost their jobs. In response, One Fair Wage launched a COVID-19 service workers' relief fund, and 270,000 workers applied. Two-thirds of tipped workers who applied reported that they faced great challenges accessing unemployment insurance, since in most states their subminimum wage was deemed too low to qualify for unemployment benefits, and in many cases their tips were not counted toward the calculation of their benefits.[16]

The mass upheaval in the restaurant industry caused by COVID-19 led to a historic turning point for the country, and for me as an organizer. Being refused unemployment insurance made many workers realize that tips were not a reliable source of income and were not even recognized as legitimate income by the government. It was a moment of awakening for millions of workers. Many had been trapped for years in subminimum-wage jobs because they thought the alternative—unemployment—was the worst thing that could possibly happen to them and their families. But when that thing happened—the massive job losses caused by the pandemic—they found themselves still standing but being penalized for having lived off tips. Many of these workers had taken great pride in working in hospitality and would have stayed in restaurants, but now they were no longer willing to accept the existing wage structure. As a result, we heard from many workers that they were leaving the industry and did not want to return until they were guaranteed a full, livable wage with tips on top.

Of course, not all food service workers left the industry. Many returned in summer 2020, when restaurants started reopening outdoor dining facilities. There, these workers faced a different kind of personal and economic crisis. In fall 2020, we surveyed thousands of workers who had returned. Many workers reported on the extremely high health risks they faced working in restaurants. In fact, one in three reported that someone in their restaurant had died from COVID-19.[17] No wonder the CDC named restaurants the most dangerous place for adults to be during the pandemic and the University of California, San Francisco, named restaurants the most dangerous place to work, ahead of hospitals, during the pandemic.[18]

Health risks weren't the only problem. An overwhelming majority of all restaurant workers surveyed reported that tips had declined significantly—not surprising, given how much sales declined during the pandemic. In addition, a majority of women reported that sexual and other forms of harassment by customers increased significantly. We worked with MacKinnon to publish research based on hundreds of women reporting comments from male customers asking servers to remove their masks so that he could judge their looks and determine tips on that basis—a phenomenon we called "maskual harassment."[19] Due to the pandemic, harassment of women food service workers escalated from unjust to a matter of life and death, with waitstaff being asked to risk their lives and the lives of their families in hopes of receiving the tips needed to help pay the household bills.

Perhaps the greatest challenge we heard about was the requirement to do much more for much less. When workers were asked to enforce COVID-19 protocols on customers—asking them to sit six feet apart, to wear masks when not eating, and to show vaccination cards—they were placed in an impossible situation. These workers were now asked to become public health marshals on top of being servers and bartenders, all for less than the minimum wage and greatly reduced tips. An overwhelming majority of workers reported that they were tipped less when they sought

to enforce these protocols. In some instances, workers were met with harassment and even violence for trying to enforce the rules.

All of these conditions were much worse for Black workers. Sixty
percent of all workers said their tips declined during the pandemic, but
70 percent of Black workers reported the same. Over 70 percent of all
workers said they were tipped less when they tried to enforce COVID-19
protocols, but over 80 percent of Black workers said the same. American
customers were not happy about being asked to follow the rules by restaurant workers—and when those workers were African Americans, the re-
sentment was even greater.[20]

Remember, too, that workers in other high-risk industries were
awarded hazard pay for their dangerous work during the pandemic—and
rightly so. But in the restaurant industry, workers were asked to take on
double duty as servers and public health marshals without even receiving
the minimum wage, and all while receiving greatly diminished tips.

FROM THE "GREAT RESIGNATION"
TO THE "GREAT REVOLUTION"

Not surprisingly, then, workers started to leave the food service industry in droves. The press did not identify the phenomenon of the "Great
Resignation" until 2021, when restaurants were allowed to fully reopen
and discovered that they did not have enough staff to do so. But we had
heard that workers were leaving the restaurant industry en masse as early
as 2020, thanks to the cascading economic and social burdens they were
being asked to bear.

To date, we have documented that 1.2 million workers have left the
restaurant industry since the start of the COVID-19 pandemic.[21] Some in
the media and elsewhere have seized on such data to feed a narrative that
workers are staying away because they are lazy and prefer to subsist on
unemployment insurance.

At One Fair Wage, we've been pushing back against this false narrative. The majority of workers we surveyed faced great challenges accessing unemployment insurance to begin with. But even more importantly,
surveys show that restaurant workers don't want to stop working. What

they want is a fair wage for the work they do. In surveys conducted during 2022, 54 percent of workers who remained in the industry reported that they were planning to leave. Seventy percent blamed their decision on low wages and tips, and over 80 percent said that the only thing that would make them return to working in restaurants would be a full minimum wage with tips on top.[22] And when we surveyed workers in states that prematurely ended unemployment insurance in a misguided attempt to push workers back into the low-paying restaurant industry, we found an even higher percentage of workers citing low wages as the reason they were leaving the industry and a full, livable wage as the only factor that would make them return.[23]

Thus, the worst staffing crisis in the history of the restaurant industry is not a "worker shortage" or a "labor shortage," as so many seek to label it. It is a wage shortage. There are plenty of people willing to work in restaurants; they are just not willing to work for poverty subminimum wages any longer. Particularly with runaway inflation in 2022, workers were making a very rational decision not to go to a job with an hourly subminimum wage that was less than the cost of a single gallon of gas. Workers were not going to spend more to get to work than they earned when they got there.

While so many employers decried these workers as "lazy," they were doing exactly what these same employers had said they could do for generations. When workers complained about low wages and terrible conditions, they were constantly told, "If you don't like it, leave." These workers listened, and left.

What has resulted in the wake of this Great Resignation strongly resembles one of the largest actions of withholding labor—in other contexts called a strike—in the history of the United States. Workers have left the industry en masse and are saying they will not return without wages going up. In response, wages have been going up in many restaurants located in the forty-three states that still permit the subminimum wage for food service workers who receive tips. No wonder workers in our organization are calling this a "Great Revolution."

In the fall of 2021, One Fair Wage began scraping job postings from restaurants off the hiring website Indeed. To date, we have tracked over five thousand restaurants in all forty-three subminimum-wage states that are offering much higher wages—in the overwhelming majority of cases, a full minimum wage or more.[24] In Dallas, Texas, where the subminimum wage for tipped workers is $2.13 an hour, we have tracked restaurants paying $25 an hour plus tips. In Cape Cod, Massachusetts, where the subminimum wage for tipped workers is just over $5 an hour, we have tracked restaurants offering $50 an hour plus tips. We are seeing wage increases from 300 to 1,000 percent in restaurants across the country.[25] Dramatic? Yes—and a reflection of the fact that food service wages have been artificially and intentionally stagnant for generations. The increases being offered by employers reflect what workers actually need to live and to continue working in restaurants: to cover the costs of getting to work, often across great distances, to pay for childcare, and then to have enough remaining to cover basic necessities.

Even more, we are starting to see some restaurants consider the notion of a living wage that actually reflects the professional skills of a hospitality worker. Just as the pandemic was a moment of epiphany for millions of workers, so, too, did we see thousands of restaurants begin to reexamine their business model. During the shutdown, we heard from hundreds of restaurant managers and owners that it was time to look at the broken system of restaurant work—a system that has resulted in one of the highest rates of employee turnover, at great cost to employers.

Many restaurant owners saw their employees struggle to enforce COVID-19 protocols on the same customers from whom they had to get tips and realized it was not a workable system. Others, having changed their business model during the pandemic to offer only takeout service, switched to paying all workers a full minimum wage with tips on top and sharing tips among all nonmanagement staff. Many found this to be a much more effective model of team building and service delivery.

Some restaurant owners told us that, after the murder of George Floyd in summer 2020, they decided to move away from the subminimum wage for tipped workers—both because it is a legacy of slavery and because it has been a proven source of racial inequity. In 2021, thousands more restaurants began to voluntarily transition to paying a full minimum wage for tipped workers in order to recruit and retain staff during the worst labor crisis in the history of the US restaurant industry.

For all these reasons, this is a moment of reinvention, renewal, and redemption for the restaurant industry—a moment that brings the possibility of a more sustainable sector finally within reach.

CODIFYING WAGE INCREASES THROUGH PERMANENT POLICY CHANGE

As the founder and leader of several organizations working to organize restaurant workers, employers, and consumers for improved wages and working conditions, I realized that we had reached a historic climax in our work. We had been advocating for a policy to raise wages and end subminimum wages in the restaurant industry for two decades. The issue is generally incredibly popular—most voters agree that workers should be paid a full minimum wage by their employers—and so, for generations, the Other NRA has thrown millions of dollars into misinformation campaigns and lobbying efforts against these wage increases. As a result, although we had fought for and won one fair wage—a full minimum wage with tips on top—in Maine, Michigan, and Washington, DC, in each instance, the restaurant industry successfully lobbied legislators to overturn democratically achieved victories.

But now we are fighting back, buoyed by the realization that everything has begun to change with the Great Revolution. In February 2022, realizing that we were in the midst of historic change and needed to support workers who were demanding permanent wage increases in order to return to the industry, we launched a campaign called "25 by 250."[26] We committed to passing legislation to raise wages and end subminimum wages in twenty-five states by 2026, when the United States will celebrate the 250th anniversary of the signing of the Declaration of Independence.

We began our efforts by collecting signatures in Michigan, where we had previously won a wage increase but had it taken away unconstitutionally. In 2018, we had collected four hundred thousand signatures to put a wage increase for all workers to $12 an hour—including tipped workers, who earned just over $3 an hour in Michigan—on the ballot. The Republican-led legislature took our measure off the ballot and made it law, but they did so with no intention of implementing it. In fact, they actually told the press that they were doing this only to suppress the vote among low-wage workers and promised to reverse the legislation after the election.[27] In July 2022, Michigan lower courts declared this action unconstitutional and stated that the original law passed—$12 plus tips—would be the law of Michigan. As of mid-2023, the issue is making its way up to the Michigan Supreme Court. If the Supreme Court agrees with the lower courts, Michigan will become the eighth one fair wage state in the United States.

As the next step, we have now collected six hundred thousand signatures to put a measure on the 2024 ballot that will raise the wage for about a million workers in Michigan to $15 an hour plus tips. We hope this measure will drive hundreds of thousands of unlikely voters who are low-wage food service workers to "vote themselves a raise."

In Washington, DC, workers and voters are similarly due a raise. The city council there had overturned a 2018 decision by 56 percent of voters to raise tipped workers to the full minimum wage, with tips on top. We collected enough signatures to put one fair wage back on the ballot in November 2022 and won with 76 percent of the vote—a dramatic increase due to increased support from restaurant workers, employers, and voters, motivated in part by the pandemic.[28]

Thousands of service workers across the country are leading the effort to advance legislation in another dozen states in 2023. We also hope to put the issue on the ballot in several more battleground states—including Ohio and Arizona—in 2024, when food service workers could be motivated to vote in a way that could impact the presidential election. Ultimately, all of these states advancing policy to codify wage increases that are already happening in the industry can build the momentum we need to pass legislation in Congress that would raise the federal minimum wage and finally end the subminimum wage for tipped workers. Our goal is to make it happen by 2025, 160 years after emancipation and just in time for our nation's 250th birthday.

KEEPING UP THE MOMENTUM

The COVID-19 pandemic has been a terrible scourge, costing more than a million Americans their lives. But the economic, social, and political impact has been more mixed. It has forced some industry sectors to finally grapple with the unfair and unsustainable practices on which their businesses were built. As a result, millions of workers in America's food service industry are no longer willing to accept poverty subminimum wages. As Ifeoma Ezimako, one of the worker leaders at One Fair Wage, has put it, "We now know our worth."[29] And this realization is having immense implications even beyond the restaurant industry, throughout the food sector and economy wide.

Because the Other NRA has played such an outsize role lobbying against wage increases—not only for restaurant workers but also for farmworkers and, ultimately, all workers—the restaurant industry has played an outsize role in setting the wage floor of the food system and the economy.[30] But now, the courage of millions of restaurant workers, particularly women and women of color, to stand up for themselves and demand their worth is driving up wages in the sector that has historically had the lowest wages of the food system and the US economy overall. Despite facing opposition from the most powerful employer trade lobby in the country, these workers are demanding more—and winning it. Let's all do what we can, as citizens and consumers, to support their efforts and keep up the momentum that is moving our country, slowly but steadily, toward a future that is more just for all.

8.

Changing the Culture of Capital to Support Regenerative Agriculture

By Lauren Manning and David LeZaks

Lauren Manning, Esq., LLM, focuses on the intersection of food, farming, and finance. She was a civil litigator before working in agrifood tech-focused venture capital for seven years. She is an adjunct professor of law in the University of Arkansas School of Law LLM program in food and agriculture law and policy, and she raised grass-fed beef, lamb, and goat in northwestern Arkansas for eight years.

David LeZaks, PhD, is an environmental scientist and financial activist whose work is centered around developing innovative mechanisms for financing the transition to agroecological farming and food systems. He completed his PhD in environment and resources and an MS in land resources at the University of Wisconsin, Madison. He is based in Madison, Wisconsin, where he is active in a number of community organizations and spends his spare time gardening and participating in a variety of silent sports.

The American farmer and rancher is one of the most storied jewels in the crown of our American identity. We still take pride in our farmers and ranchers, honoring them as one of the longest threads in the fabric of our culture through everything from Super Bowl commercials to bumper stickers quipping, "If you ate today, thank a farmer." This is understandable. After all, agriculture is one of the few industries that humans *must* engage in if we want to survive. We can live without many of the creature comforts we enjoy in modern society, but we cannot live without food— or without farmers and ranchers.

Unfortunately for Thomas Jefferson, our modern attempts to idolize farmers and ranchers ring hollow when viewed against the unwarranted financial burdens we've laid on their backs.

Make no mistake, producing food is a lucrative venture for a few farmers: the biggest and best aligned with corporate agribusiness interests. Even these farmers, however, are often beholden to those corporate interests or trapped inside production contracts that render them modern-day serfs on land they were lucky to inherit or took on massive debt to own. As we'll explain, what they grow and how they grow it is dictated largely by the financial safety net our government policymakers have offered them, favoring commodity crops like soy and corn (with much of the latter ending up in fuel tanks rather than on dinner tables).

And despite that safety net, for most producers, farming is a brutal exercise in economic survival. As a result, 96 percent of farm households receive off-farm income, and those earnings provide 82 percent of total income for all family farms (2019 figures).[1] The people we are paying to grow our food cannot even afford to buy their own groceries with farming income alone. They break their bodies to feed us, yet most of them cannot afford health insurance or the out-of-pocket costs of seeking medical treatment. For many farmers, sending their kids to college, squirreling away retirement funds, or taking a real vacation are sheer fantasies.

American farmers and ranchers frequently put their land up as collateral to take out loans to buy new equipment or simply to cover the astronomical cost of the inputs they must buy—things like seed, feed, and fertilizer. If you're a farmer who has bravely stepped off the commodity treadmill and forfeited your access to government programs like crop insurance and subsidy payments with the hopes of growing food in a different way, you are one severe weather event or supply-chain crisis away from losing your livelihood and your home.

But the realities of today's food production industry have left our producers carrying a disproportionate share of the risk inherent in the business, despite the fact that absolutely nothing happens in this food system without them. The farmers and ranchers who are doing the most important part of the work often have the most to lose.

We have a choice, however. As a society, we can choose to take responsibility for the economic environment that we've created for farmers and ranchers, and we can help shoulder the burden alongside them. We can choose to pull farmers and ranchers from the economic abyss into which we've let them fall and put them back where they belong, at the heart of our society. In that role, they can remind us of an essential truth: We humans belong to the Earth. It does not belong to us.

Fortunately, a new economic path for farmers is already being forged by some members of the financial community. Their innovations are providing examples of how a different approach to agricultural finance is possible—one that shares risk more equitably and makes it possible for farmers to produce food in sustainable, healthy ways while earning a fair share of the profits.

We'll start with a quick lesson on how we got to where we are. Then we'll outline the positive changes that are starting to emerge and leave you with some ways you can help accelerate those changes.

HOW WE GOT HERE: A SHORT HISTORY

As humanity has evolved from being hunter-gatherers to agrarians to primarily urbanites, we've off-loaded food production to a relative handful of people. Agriculture, food, and related industries contributed over $1 trillion to the US gross domestic product in 2020, yet only 10.3 percent of Americans are employed in the industry, according to USDA Economic Research Service (ERS) data, and only 1.4 percent of Americans are employed on farms.[2]

This is a dramatic change from the circumstances just one hundred years ago. In 1930, 25 percent of the US population, or roughly thirty million people, lived on farms or ranches, according to USDA Census of Agriculture Historical Archive data.

This change has freed up the vast majority of our workforce to engage in other activities, including medicine, science, education, and art. But this outsourcing has resulted in a societal loss of understanding about how food is grown, who grows it, what it costs to grow, and how it gets transported, and processed, and distributed.

Some would applaud the fact that American supermarkets are overflowing with cheap, abundant food, including seasonal foods available year-round. After all, isn't it the goal of our food system to feed people as cheaply and efficiently as we can? Yes, affordable food is an important social good. But it's equally important to look closely at how that food is being produced and the long-term impact on all of us.

Let's start by acknowledging the reality that, for a large segment of American households, relatively cheap food is still a major part of the household budget. In 2021, households in the lowest income quintile spent an average of $4,875 on food, reflecting nearly one-third of total income.[3] Affordable, abundantly available food may be an indulgence for the affluent, but it is a necessity for the financially less fortunate.

Unfortunately, there is no such thing as a free lunch. The true costs of getting low-cost food onto the tables of American families has to fall on someone else's balance sheet. Some call these the hidden costs of food. They include the costs of environmental damage, inequities in the distribution of taxpayer-funded farm support programs, the rising cost of agricultural inputs like fertilizer, and the costs of treating illnesses related to unhealthy diets, such as heart disease and diabetes.[4]

In fact, according to a Rockefeller Foundation analysis, while national food expenditure totals roughly $1.1 trillion per year, the true cost of producing that food is three times higher at $3.2 trillion per year, with the overwhelming majority of those costs contributing to the degradation of human health.[5] Most consumers are insulated from these economic realities, but, in their role as taxpayers, they ultimately help to pay for these hidden costs. In the process, the farmers themselves usually end up with little to show for their work.

A number of factors have led us to this era of cheap, abundantly available food. In response to the dust bowl of the 1930s, policymakers enacted programs that paid farmers to let land lie fallow when production soared and prices plummeted. The goal was to protect producer incomes. When prices became too high, subsidy payments ceased and production resumed to ensure a sufficient food supply and ease the hit on consumers' wallets. Fallowing land also eased some of the ecological pressure that had triggered the dust bowl's devastation.[6] The system worked reasonably well for a time.

In 1971, President Richard Nixon appointed Earl Butz as secretary of agriculture. Butz promptly altered many New Deal–era programs on the basis that they imposed unnecessary restraints on producers' autonomy, preventing them from planting as much as they wanted whenever they wanted. With an eye toward capitalizing on foreign trade opportunities, Butz promised that any surplus would be shipped overseas to prevent domestic prices from dropping too low.

Butz secured a multiyear sale of the US grain reserve to Soviet Russia. But drought befell the Midwest, sending grain prices soaring. Producers saw an opportunity to capitalize on the high prices by planting as much as they could. Butz supported this tactic, infamously encouraging them to "plant fence row to fence row" and to "get big or get out."

To increase their production capacity, producers took on significant debt to acquire more land, machinery, and inputs. Between 1970 and 1980, farm-sector debt soared from roughly $300 billion to $500 billion.[7] At the dawn of the 1980s, things took a disastrous turn for producers who were leveraged to the hilt. Interest rates skyrocketed, the Federal Reserve shifted policies to tamp down inflation, President Jimmy Carter enacted a grain embargo on Russia in response to its invasion of Afghanistan, overproduction sent prices on a downward spiral, and tens of thousands of farms went under.

The ripple effects of the 1980s farm financial crisis were severe, forcing input manufacturers and equipment makers like John Deere to slash thousands of jobs. As farms disappeared, rural communities faded with them. Suicides among producers peaked at roughly fifty-eight per one hundred thousand.[8]

The federal farm policies of the "get big or get out" era did tremendous damage to farmers. But they also caused harmful effects in the entire web of our food system. The major grain companies saw an opportunity to take advantage of the enormous grain supply by launching new markets around high-fructose corn syrup and ethanol—markets that subsequently distorted the eating habits of millions of Americans and diverted vast amounts of farmland to the production of fuel for automobiles. Meanwhile, the cattle industry capitalized on the abundance of cheap animal feed, encouraging Americans to eat even more meat.

The agricultural biotechnology industry also capitalized on the "fence row to fence row" mentality. Monsanto introduced Roundup Ready soybeans in 1989 and Roundup Ready corn in 1998—genetically modified varieties designed to survive the heavy use of chemical pesticides. By 2020, more than 90 percent of corn, upland cotton, and soybeans would be produced with genetically engineered seeds.[9] Agrochemical company Bayer purchased Monsanto for a staggering $63 billion in 2018.

The hidden costs of our food production system also include long-term environmental degradation. Here's just one example: The USDA Natural Resources Conservation Service estimates that Iowa alone has lost 6.8 inches of topsoil since 1850, which harms agriculture yields to the tune of ten fewer bushels of corn per acre. Meanwhile, the Iowa Daily Erosion Project estimates that soil loss is costing Iowa corn growers as much as twenty-nine bushels per acre on highly erodible land. Overall, Iowa's ongoing soil loss has cost producers in the state an estimated $1 billion in revenue.[10]

Current government policy has not only ignored this sleeping giant of a problem; it's condoned it. The USDA maintains a metric called T, which refers to the amount of soil loss that is tolerable on a farm.[11] Producers have largely offset yield reductions due to soil loss and poor soil health by applying more nitrogen fertilizer and choosing different biotechnology-derived seed varieties—short-term fixes that do not solve the underlying problem.

The long-term impact of the food policies launched in the 1970s and 1980s is still with us today. Most current agricultural finance policies still favor high yields and maximum efficiency with little regard for the impact that aggressive production strategies and a prevalence of commodity-based monocultures have on the environment, human health, social equity, and—as the COVID-19 pandemic demonstrated—food security.

Follow your dollars around the food system and you will soon discover the DNA of this policy platform. It's revealed through financial practices such as extractive loan-underwriting policies driving soil loss, the valuation of agricultural land for development eating away at rural communities, government farm safety-net programs that mainly help the rich get richer, dismal wages and conditions for supply-chain workers, and lobbying's heavy-handed influence in perpetuating these economic illnesses for economic gain.

The yield-hungry policies put into place during the Butz era have led to ecological decline, low incomes for producers, less nutritious food, and a vulnerable supply chain. This policy environment has bred a culture of private capital with a similarly myopic hunger for profits and growth regardless of the cost.

But policy can be changed and the entire industry along with it. A growing chorus of stakeholders are championing new approaches to growing food that prioritize noneconomic outcomes like environmental regeneration and a move toward non-extractive agriculture, where the returns to the lender or investor do not exceed the wealth created by the farmer.[12]

Without adequate financial support and a favorable policy environment, however, these movements have little hope of scaling across the 911 million acres of farmland in the United States.[13] Changing federal policy is a herculean task made thorny by lobbyists with substantial corporate-backed war chests. And until policy changes, private capital has little incentive to modify its posture from yield-focused capital providers to partners who could shoulder the risk of transitioning to a more regenerative agriculture alongside producers. Private capital may not be ready to make this transition until it has demonstrable proof that there is a different way to finance the business of feeding ourselves.

Under the circumstances, philanthropy has a unique opportunity to reveal a different path, along with any members of the private capital space who are willing to buck against conventional culture and demand something different from their peers.

Designing a new financial ecosystem for agriculture that is willing to shoulder the yoke of risk, economic uncertainty, brutally hard labor, supply-chain vulnerability, and an increasingly volatile climate with our farmers and ranchers will not be an easy task. But if we continue down our current path of extracting everything we can from the land and the people who grow our food without any consideration for the cost, we may not make it out of the twenty-first century.

HOW THE PUBLIC SECTOR HELPS TO DRIVE UP THE HIDDEN COSTS OF FOOD

Farms and ranches are largely private businesses that face all of the same challenges as other businesses when it comes to cash flow projections, managing input costs, and achieving financial margins that leave them with enough cash to buy their own groceries, purchase health insurance, send their kids to college, and maybe even take a vacation. However, examination of the average farmer's financial realities reveals a challenging landscape where decisions about what to plant, when to plant, how to plant, and where to sell it are constrained by crushing economic realities.

The business of growing food often does not pay enough for producers to support their own families. In 2017, ERS data showed that producers receive a paltry 7.8 cents for every dollar spent on food.[14] This explains why, as we've noted, the vast majority of farming households must rely on members employed in some off-farm capacity.

Farm safety-net programs like crop insurance and crop subsidies make up the bulk of taxpayer-funded support payments to producers.[15] They are managed under an omnibus piece of legislation called the Farm Bill that is reauthorized every five years, along with other programs like ad hoc disaster aid, voluntary conservation support, and food assistance programs. There are roughly nineteen crop insurance programs covering 124 crops including rangeland, pasture, and livestock. However, four commodities—corn, soybeans, wheat, and cotton—garner the overwhelming bulk of the payments.

Taxpayers fund roughly two-thirds of the premiums for crop insurance, which are administered by private crop insurance companies—also supported by taxpayers. While premium subsidies totaled nearly $890 million in 1995, they ballooned to $6.3 billion in 2020. The administrative costs for crop insurance exceeded $1 billion annually between 2007 and 2018.[16]

A second major component of the federal farm safety net is crop subsidies, which are intended to help producers manage fluctuations in prices and yields due to factors like weather and market volatility. Over the past decade, subsidy payments have cost taxpayers roughly $16 billion per year, with much of that capital being highly concentrated around a handful of commodities like corn, soy, wheat, cotton, and rice.[17]

Producers have two options for participating in subsidies. First, agriculture risk coverage (ARC) provides payments to farmers if their revenue per acre or their county's average revenue per acre drops below a guaranteed level.

Second, price loss coverage (PLC) pays farmers on the basis of the national average price of a crop compared to the crop's reference price as established by Congress in the Farm Bill. The reference prices, which are set intentionally high, are adjusted annually based on historical market conditions. Payments are made according to the number of "base acres" enrolled in the program, which represent a farm's crop-specific acres of eligible commodities. However, payments are made regardless of what is actually planted on the farm.

Producers enrolled in PLC receive a payment that is equal to 85 percent of their base acreage times the difference between the reference price and the effective price, times the PLC payment yield (the established yield of the farm) for the covered commodity.[18] The 2018 Farm Bill gave producers a one-time opportunity to update their payment yield, effective for the 2020–2021 season.

On the surface, these programs appear to offer positive incentives that free producers from worrying about risks like weather or market volatility and to simply grow as much as they can. But negative consequences await farmers who choose to step off that treadmill and endeavor to grow something different using different methodologies and with different outcomes in mind. They must face the economic impacts of any disaster almost entirely alone. The result is a sense of financial dependency that discourages innovation and experimentation.

What's more, nearly a century after Roosevelt's New Deal, many argue that farm safety-net programs have slipped out of alignment with their original goal: ensuring the survival of small-scale, independent producers.

Instead, over the past decades, rapid consolidation in agriculture, technological advancements, and a shift toward farm specialization have widened the gap between the smallest and the largest operations—with help from the federal subsidy programs.

Small farms with gross cash income under $250,000 make up 91 percent of American farms, according to the USDA. Yet the largest farms receive the overwhelming majority of farm safety-net payments.[19] During 2016, 17 percent of subsidy payments went to the top 1 percent of farms, while 60 percent went to the top 10 percent.[20] A number of politicians and wealthy businesspeople are subsidy recipients, including NBA Timberwolves owner and *Forbes* 400 member Glen Taylor, who received $116,502 in subsidies during 2017.[21] Some have argued that the farm subsidies are nothing more than a form of welfare for the wealthy and call into question how many operations would immediately become unprofitable if they stopped receiving crop insurance and subsidies.

Moreover, loopholes in the programs invite rampant malfeasance. Although there is a $125,000 annual payment limit for subsidies, a farming operation can list an unlimited number of partners eligible to receive the maximum amount each year if they demonstrate that the individual is "actively engaged in farming" by showing a significant contribution of capital, equipment, or land and involvement with labor or management.[22]

A 2021 study on the impact of subsidies on farmer behavior also concluded that subsidies often disincentivize producers from risk-mitigation efforts. If crop insurance and subsidy payments neutralize negative impacts from low prices or poor yields, a farmer has less incentive to take proactive steps against those outcomes—for example, by paying for cover crops to boost soil health, diversifying rotations, or repairing degraded riparian buffers to prevent nutrient runoff.[23] As climate change creates increasingly unpredictable weather patterns and devastating weather events like derechos (destructive wind storms), floods, and droughts, policies that incentivize practices conducive soil health and diversification in lieu of yield alone are critical to protecting our food supply.

The promise of potentially lucrative farm safety-net program payments has also contributed in part to a farmland tenure crisis. Today, 60 percent of farmland is owner operated, meaning the person who owns the land is also managing it. The remaining 40 percent is rented out by so-called non-operator landowners. That figure is even higher in some states such as Illinois, where 77 percent of farmland is rented.[24] Although the ability to lease farmland has benefits for those who do not want to expand their land base or who cannot afford to buy land, there are dire consequences too. Producers who rent land have little economic incentive to invest time or money in conservation or land rehabilitation, knowing that they may lose that lease to the higher bidder next year. In many cases, the leases are controlled by individuals who inherited the multigenerational family land and who do not even live in the same community. The promise of passive income from an annual rent check is an easier pill to swallow than the guilt of selling the family land to the producer who manages it.[25]

While taxpayers pay billions of dollars for the farm safety-net programs, the health-care industry also picks up a heavy share of our food system's hidden costs. The cost of treating chronic diseases related to unhealthy diets, such as heart disease and obesity, makes up 90 percent of the $4.1 trillion in annual health-care expenditures in the United States, according to the CDC.[26] Diabetes costs roughly $327 billion, while obesity, which affects 20 percent of children and nearly half of adults, costs $173 billion annually.

While taxpayers pay these public costs, corporate interests play a heavy hand in crafting the health-care system that manages them. There are countless examples. Here's just one: In 2016, a study showed that, decades before, the sugar industry had sponsored research at Harvard that concluded—erroneously—that sugar does not contribute to heart disease. In 1967, Harvard scientists had published a study reaching that very conclusion (and claiming that cutting out fat was the best way to reduce heart disease) without disclosing the source of the study's funding.[27]

Rather than encouraging Americans to adopt healthier eating habits, US industry has a vested interest in producing and selling medications aimed at alleviating the diseases caused by unhealthy foods. Accordingly, between 1999 and 2018, the pharmaceutical industry spent roughly $4.7 billion on lobbying while making $414 million in contributions to political campaigns. It shelled out another $877 million in contributions to state-based candidates. Most of these contributions were targeted toward senior legislators responsible for drafting health-care laws or committees that oversee referenda on drug pricing and regulation.[28]

Meanwhile, the nutritional integrity of our food—that is, the levels of essential nutrients such as protein and vitamins—has declined dramatically since the 1950s. A 2004 study attributed this decline to the varieties of foodstuffs that producers choose to grow, often prioritizing yield over nutritional value.[29] Of course, the decline also coincides with the "get big or get out" policy era and the ubiquitous adoption of new farming technologies like nitrogen fertilizer and glyphosate—again prioritizing yield over everything else.

CAN VENTURE CAPITAL AND PRIVATE EQUITY FINANCE FOOD SYSTEM REFORM?

Federal programs and policy set the tone for how agriculture and food production are structured throughout the United States. However, the bulk of the finance that supports producers comes from the private sector, which means that the rules governing private finance play a major role in shaping our food system.

In fact, an increasing number of private capital providers are engaging with the food system for a diverse set of reasons. Venture capital and private equity firms are pouring staggering amounts of capital into food and agriculture, investing $51.7 billion alone in 2021, nearly doubling from 2020.[30]

As a result, those of us who want to see reform of the US food system to produce greater fairness for farmers and healthier, more sustainable diets for families have been hoping that private finance can help lead the way. But how much of the growing amount of private capital will actually end up directly improving farming practices, farmer livelihoods, and ecological outcomes is yet unknown.

The private sector offers a number of ways to access capital for producers, but these options are not without their complexities. A substantial majority of that capital goes to upstream entities like e-grocery start-ups, which captured $18.5 billion. What's more, just as government economic policy has favored high yields at all costs in agriculture, high returns reign supreme throughout most private sector finance. These venture capital and private equity firms set high financial expectations—for example, ten times return on their investment. And although investment managers and philanthropic-foundation program officers decide how to deploy capital, they are often constrained by the expectations of the individuals who seed the funds that they manage, including high-net-worth individuals, institutional investors, and corporations.

The high returns demanded by these influential parties make it hard for private capital to be a driver of food system reform. For example, venture capital and private equity may help commercialize technologies that produce food with fewer inputs or develop a new class of bio-based inputs to replace existing agrochemicals. However, these pools of capital typically lack the patience and flexibility that producers need when transitioning to regenerative agriculture, which values healthy soil, biodiversity, water quality, safe working conditions, nutritional quality, and food security rather than the biggest financial returns possible. Transitions come with high up-front capital costs and risks associated with changing what is grown, how it is grown, and where it is sold. In essence, many producers transitioning to regenerative agriculture are forced to create their own independent supply chains while becoming marketing, branding, and distribution managers. This can result in longer timelines to profitability and slow growth.

There are other problems with relying on private capital to drive food system reform. Many in the venture and private equity spaces tout carbon credit markets as a way to pay producers for ecosystem services like soil restoration and habitat preservation, but critics express concerns about whether the methodologies used to measure levels of carbon sequestration as well as its permanence are too thin to provide a foundation for carbon credit markets. Some emerging carbon credit markets are basing payments on statistical models that estimate how much carbon can be sequestered using practices like cover crops and no-till drilling, to avoid the high cost and labor intensiveness of ground sampling. Critics are also concerned that focusing solely on carbon sequestration will preclude other important factors from entering the dialogue, such as the continued use of synthetic fertilizer and agrochemicals or farmworker safety.

For all these reasons, it seems unlikely that venture capital and private equity investors will provide the resources farmers need to make the US food system healthier, fairer, and more sustainable.

WHAT ABOUT LENDERS?

Debt capital is the predominant source of financing for food producers. The Farm Credit Administration is an independent financial regulatory agency that oversees the Farm Credit System (FCS), a nationwide network of lending institutions that serve farmers, ranchers, agricultural cooperatives, and other eligible borrowers. It also oversees the Federal Agricultural Mortgage Corporation, or Farmer Mac, which provides a secondary market for agricultural real estate mortgage loans, rural housing loans, and rural cooperative loans.

The largest single private lender into agriculture is the FCS. As of 2021, it held over $200 billion of farm loans, representing almost 45 percent of agricultural debt. The FCS is a government-sponsored enterprise, which means it has certain responsibilities and mandates as determined by Congress. It also has both tax advantages and a lower cost of capital than other private banks, due to the implicit guarantee that the government will bail out the FCS in the event it fails.[31]

There's a big problem with the FCS, however. Although the agency's policies have recently become more sophisticated, especially when compared to non-savvy lenders, its underwriting practices are still largely steeped in the ethos of the Earl Butz era. Loan applications are evaluated solely based on whether a producer will yield enough profit to repay the loan over the required period. Soil is not viewed as a depreciating asset, and noneconomic outcomes like water quality, reduced nutrient runoff, and wildlife are not reflected on the balance sheet.

As the top loan provider to some of the largest pork conglomerates in the United States, FCS perpetuates the kinds of harmful practices that characterize the entire food system. For example, in addition to the climate impacts of high-density confinement livestock operations, such as water contamination and nutrient runoff, pork production is responsible for using 27 percent of all medically important antibiotics sold in the

United States, perpetuating antimicrobial resistance and reducing the effi-
cacy of antibiotics for both humans and animals.[32] FCS loan policies have
done nothing to alter these practices. And while FCS has recently formed
a climate change task force, it is years behind the rest of the government
and parts of the private sector in incorporating climate change into its
lending operations, underwriting practices, and risk calculations for the
financial products it sells on the public markets.[33]

Thus, considerable changes are needed to align this government-
sponsored enterprise with the realities of the challenges that agriculture
faces, including climate change, soil loss, and human health. Bringing
these changes to fruition will require pressure from multiple stakeholders—
policymakers, investors, farmers, and consumers—who all share an interest
in creating a resilient, self-sustaining rural economy.

Producers can also access loans through the USDA Farm Service
Agency (FSA), which provides direct and guaranteed farmland loans and
operating loans to producers who cannot obtain commercial credit from a
bank. Known as "the bank of last resort," FSA provides conventional agricultural lenders with up to a 95 percent guarantee of the principal loan
amount.

Producers wanting to transition to regenerative agriculture face many
of the same challenges they experience at FCS when attempting to access
capital through FSA, including a lack of understanding about regenerative practices and a tendency to prefer conventional methods due to their
familiarity. FSA loan officers have to *want* to learn more about alternative markets in order to understand a producer's proposed business plan.
They also need to recognize that short-term changes in farm management
practices can have long-term impacts on soil health and the land's overall
productivity, and that these realities should be reflected in the agency's
lending policies.

As it stands, a producer largely has the burden of proving to FSA that
the alternative practices or markets they wish to adopt will provide sufficient cash flow. This means the producer must have three years of financial
data to show that they are a good investment without receiving up-front
capital from FSA to obtain that data. By contrast, a producer wishing to
pursue conventional practices would most likely receive the benefit of the

doubt that those practices and markets will yield enough cash flow to pay back the loan.

BEYOND THE NUMBERS: NEW WAYS OF EVALUATING FARM INVESTMENTS

Several studies and datasets have shown that conservation practices are profitable and can even provide producers with better margins compared to conventional production systems.[34] Nonetheless, the actual or perceived risks of transitioning to regenerative agriculture have kept many private capital providers on the sidelines. Learning to produce food differently can pose a steep learning curve, and there is no guarantee that such producers will find markets willing to provide them with a premium payment for goods based on the noneconomic benefits they have cultivated (soil health, water quality, and so on). In the short term, a transition to regenerative agriculture may produce lower yields compared to conventional practices, souring a conventional capital provider's opinion of whether they will make their money back on top of the necessary returns.

For these reasons, although some producers are making it work with conventional debt capital, the asset class is largely out of touch with regenerative agriculture. Currently, most conventional banks fail to consider nonfinancial outcomes when engaging in loan underwriting practices. What doesn't make it on the underwriter's balance sheet is an assessment of ecological impact, resulting water quality, biodiversity levels, or reduced nutrient runoff—all factors that have long-term economic consequences.

Imagine how farm finance might be different if lenders redesigned their policies to support the values inherent in regenerative agriculture. Recall the massive levels of soil damage condoned by the current policies of government agencies as well as private lenders. Then consider how an underwriter's opinion of a potential loan deal might change if soil were counted as a depreciating asset on the balance sheet. Or if they knew that conventional farming methodologies were stripping soil away at a rate that would make that land unfarmable in fifty years? The result might be that a loan applicant promising to apply regenerative farming principles would suddenly be more attractive than a conventional farmer—precisely the opposite of current lending practice.

Unfortunately, most conventional lenders lack the incentive and acumen to attune their lending practices to regenerative agriculture's aims. Stepping off of the farm safety-net payment treadmill and converting an existing farming operation to something different—organic certified, diversified, regenerative—is reckless in many lenders' eyes. Why leave the comfort of the conventional playbook to navigate unknowns when the market is rarely willing to pay for benefits like soil restoration, wildlife habitat, and a reduced dependence on agrochemicals?

Culture comes to bear in this dynamic too. For many farmers, agriculture is more than just a job. It is a way of life and a cultural identity, particularly for individuals with a deep family legacy tied to a particular piece of land or way of life. As a result, some producers and other individuals who work in the agriculture industry view regenerative agriculture as an implied attack on conventional ways of producing food. In their view, agriculture isn't broken, and any attempts to fix it are often received as antagonistic. Thus, a capital provider in a small town reviewing a loan application for a first-generation producer wanting to raise grass-finished beef may take a biased approach in assessing the creditworthiness of this borrower. A third-generation cattle rancher wanting to shift to pastured lamb may face scrutiny and shame or even risk being ostracized from his community.

CHANGING THE CULTURE OF AGRICULTURAL CAPITAL

As you can see, there's a strong connection between how farms are financed and the kinds of agricultural practices farmers are able to employ. If we are to transition from today's broken food production system to a more regenerative style of agriculture, the farm finance system must evolve as well. We cannot expect producers to shoulder the burden and risks of transformation through bootstrapping or working a full-time off-farm job to foot the bill. At a very high level, the solution to this financial conundrum is convincing all capital providers—heads of government agencies, private investors and lenders, philanthropists, and more—to start viewing themselves as capital *partners*. It also lies in creating a financial safety net akin to the one conventional food producers currently access.

One way to jump-start this change is by leveraging the power of philanthropic capital. This category of finance encompasses donated money that is deployed in a variety of forms including grants, recoverable grants, and program-related investments (PRIs).

A grant is essentially a gift, while a recoverable grant is a grant that can be returned to the donor if the recipient achieves their objectives. A PRI is an investment made using the foundation's capital that is reserved for philanthropic purposes, like mitigating climate change or addressing social injustice. These investments are made with below market-rate return expectations.

The existence of these forms of capital offers a massive opportunity for philanthropists to demonstrate a new culture of agricultural investment that is grounded in partnership rather than pure profit. [1]

Current philanthropic strategies largely involve deploying capital to intermediaries: other nonprofit entities and charities that carry out the actual day-to-day work that the philanthropists want to see done. For example, a foundation geared toward addressing climate change may provide financial support to a number of programs working in areas such as energy, clean water, or food and agriculture.

PRIs offer a promising opportunity for philanthropic organizations to create a safety net for producers wanting to transition to regenerative agriculture or to expand their existing regenerative operations. Unlike grants, which are one-time allocations of capital, PRIs can generate returns that maintain the fund indefinitely. Philanthropists and foundation directors can also set their own expectations around returns, timing, and the nonfinancial outcomes they hope to produce. Unlike private equity and venture capital funds, foundations have the freedom to offer producers patient, flexible, and non-extractive capital.

Already, a host of philanthropic entities are rallying to this cause. Funders for Regenerative Agriculture (FORA) is a five-year initiative sparking affiliations among multiple funder networks to dramatically increase the adoption of regenerative agriculture. Chief among FORA's principles is a desire to take more risk and to embrace funding approaches that are commensurate with the challenges facing food producers interested in transitioning to regenerative agriculture. Because philanthropic capital has more flexibility about its expectations and capital-deployment strategies, it is more capable of shouldering the risks of transitioning than existing public and private finance asset classes.

Another group proving that patient, flexible, and non-extractive capital can create profitable enterprises that produce a multitude of non-economic benefits is the Rural Beacon Initiative (RBI), which seeks to advance community ownership in a new clean energy economy through strategic leveraging of relationship expertise and innovative technology. Through a partnership with the Alliance of Native Seedkeepers, the RBI team is housing a fully sustainable "seed sanctuary" that serves both Black and Indigenous communities in the cultivation and preservation of ancestral seeds as well as present-day culturally relevant foods. (The team uses the word *sanctuary* rather than the more familiar *bank* to reflect the belief that a seed is a living, sacred being that is to be revered.)

Mad Agriculture is an organization working to launch patient, flexible, and non-extractive capital sources for producers. Inspired by *The Mad Farmer Poems* of Wendell Berry, it offers radically different financing terms for producers wanting to transition to organic production, including transition loans, down-payment assistance, and succession or farm-transfer loans, as well as classic capital forms like operating loans, mortgages, equipment loans, and infrastructure loans. Perhaps most important, Mad Ag acts as a partner in the process, advising on business design, market development, diversification opportunities, and more.

Mad Ag is also tackling one of the biggest challenges facing our transition to a more regenerative agriculture: creating a new supply chain. Currently, a producer using regenerative practices will fetch conventional prices in the existing market. To compensate a producer for the noneconomic outcomes that their regenerative practices produce, new market opportunities are critical. Ideally, such new markets can also provide mission-aligned partnership and flexibility such as whole carcass utilization, byproduct up-cycling, and consumer education. Mad Ag's Mad Markets platform aims to connect farmers to mission-aligned buyers while promoting adoption of regenerative agriculture in the broader marketplace.

Mad Ag is not alone in its endeavors. Steward provides flexible loans to "human-scale" farms, ranches, fisheries, and food producers engaging in regenerative agriculture. It does this by providing qualified lenders with the opportunity to purchase loan participations.

Dirt Capital Partners invests in farmland in partnership with regenerative producers throughout the Northeast to assist with relocation, expansion, restructuring, and engaging with land trusts to keep agricultural land out of development. Iroquois Valley Farmland REIT is an organic farmland finance company that helps producers secure long-term land access and working capital through leases, mortgages, and operating lines of credit.

Important work is also being done to bring greater social equity to agriculture, which has a history of financial discrimination against non-white producers. Potlikker Capital is a farm-community governed, charitable integrated capital fund created to holistically serve Black, Indigenous, and people of color (BIPOC) producers. This involves providing capital to stabilize and grow existing businesses, access to higher-value markets, access to cooperative ownership across the food value chain, educational opportunities, and assistance with adopting regenerative agriculture. It also hosts an agroecology and agribusiness internship program to cultivate the next generation of BIPOC producer entrepreneurs. The integrated capital approach it takes involves a combination of non-extractive investments, zero- and low-interest loans, and grants and recoverable grants. Similarly, Akiptan, a Native-focused community development financial institution, is working to transform Native agriculture and food economies by delivering creative capital, leading paradigm changes, and enhancing producer prosperity.

There are more examples of lenders and investors working to build a new financial ecosystem for agriculture. It takes courage to tackle this work. Pioneers in this field face criticism, mockery, and even shunning from their peers if they fail (and sometimes even when they succeed). But their work is critical to solving not just the economic crisis that producers face but also the climate crisis and the human health crisis that our entire species faces.

HOW YOU CAN HELP, ONE DOLLAR AT A TIME

Consumers who support the effort to reshape our food system often feel helpless. They are told to "vote with their dollars" in the hope that their purchases will shape food manufacturers' practices. But as greenwashing runs rampant and major consumer-goods corporations launch splashy initiatives promising reductions in greenhouse gas emissions, it's hard to say for certain whether this economic voting power can make a significant impact across the food web.

On the other hand, you are more connected to the financial ecosystem than you probably think. Like our food system, finance is a web that touches countless points and is impacted by ripple effects. Take a look at your bank, the entity that services your mortgage, your life insurance policy, your credit card account, or your retirement savings. These institutions manage your money by investing it in public markets. Those investments often involve oil and gas, deforestation, and agrochemicals. You can begin to make a difference by consciously allocating some of your own funds to financial institutions that are supporting new and better forms of food production. Ask questions about how your money is managed and where it is invested. If you are unhappy with the answers, make a fuss. If all else fails and you are able, move your money to an entity with mission-aligned values.

Complicated? A little. But tools are emerging to help people trace their dollars and understand more about where they might be deployed in the markets, such as the "Invest Your Values" tool from As You Sow.[35] There are other resources, like Money Transforms, to help you learn more about this process.[36] More and more, financial service providers are emerging with ambitions that involve more than just making big returns. Especially if you are a high-net-worth individual, find an investment manager who understands how you'd like to see your money go to work.

The ballot box offers another opportunity to have a hand in shaping our society's agriculture policies. Are there local policies in your community that can be modified to support regenerative farming practices? For example, could a rezoning measure provide pasture-based livestock producers with a chance to purchase an old industrial building and convert it into a local, cooperatively owned processing plant? Can your community provide support for local farmers' markets, particularly those that operate year-round? Yes, these are small-scale changes whose impact will take time to be felt. But in the long run, consistently hitting singles can be more productive that struggling to hit the occasional home run.

The changes we need—financially and in other ways—to create a more regenerative agricultural system may seem daunting. But they don't have to happen overnight, and they shouldn't. Long-term, sustainable change is the goal, and such change is driven just as much by ordinary citizens as it is by philanthropists, policymakers, investors, and farmers. Join us in the effort to create a financial ecosystem that allows regenerative agriculture to thrive—one dollar at a time.

9.

A Human Rights Revolution Born in the Fields of Florida

By the Coalition of Immokalee Workers

The Coalition of Immokalee Workers (CIW) is a worker-based human rights organization based in Immokalee, Florida, committed to improving working conditions through enforceable human rights protections within supply chains. Internationally recognized for its achievements in the field of corporate accountability—with a particular focus on the fight against forced labor and gender-based violence at work—the CIW is built on a foundation of farmworker community organizing reinforced by a national consumer network. The CIW's work encompasses three broad and overlapping spheres: (1) the Fair Food Program (FFP); (2) the Anti-Slavery Program; and (3) the Campaign for Fair Food. In keeping with the CIW's organizational structure and tradition, this chapter is being credited to the entire team rather than to any individual, since it would not have been possible without the contributions of all.

In 1993, farmworkers rose up in the dusty streets of a dirt-poor Florida town by the strange name of Immokalee and declared a general strike. In a shock to the sleepy farming community unlike anything in its history, thousands of workers occupied the massive parking lot at the center of town—where normally labor buses would pick up workers before dawn each day to ferry them out to the fields—and demanded an end to the systematic violation of their fundamental human rights.

The Coalition of Immokalee Workers was born in those streets. To-day, thirty years later, through the unrelenting struggle and sacrifice of tens of thousands of workers and their consumer allies, the CIW's successful efforts have transformed an industry once dubbed "ground zero for modern-day slavery" by federal prosecutors into what one expert called "the best working environment in American agriculture" on the front page of the *New York Times*.

The product of that struggle—and the means by which CIW achieved that transformation, the Fair Food Program—was forged by farmworkers themselves. A diverse coalition of Mexican, Guatemalan, and Haitian immigrants grew tired of being beaten, raped, and robbed by their bosses and came together behind a vision of a more modern, more humane industry. Refusing to be trapped by poverty and powerlessness, they organized in their own community and educated consumers across the country, faced off against some of the world's largest corporations, and won. In the process, those same farmworkers created a new model for the protection of human rights in corporate supply chains globally, a model known today as worker-driven social responsibility (WSR).

WSR harnesses the unprecedented buying power of the world's billion-dollar brands, from McDonald's to Walmart, and reverses the impact of that power, improving wages and working conditions in their suppliers' operations rather than impoverishing the lives of the millions of workers around the globe who put food on our tables and clothes on our backs. The general strike that morning nearly thirty years ago in Immokalee—a protest that barely even made the local papers—ended up being the first volley in what has become a growing revolution in the world of business and human rights.

THE FAIR FOOD PROGRAM

The Fair Food Program is a unique human rights protection initiative, developed by workers for workers in order to safeguard their basic human rights, and backed by major corporate buyers who agree to only source produce from farms that comply with the program's code of conduct. On FFP farms, workers are protected from abuses like forced labor, sexual harassment, and wage theft through several interlacing mechanisms designed to monitor and enforce the program's standards, including: worker-to-worker education on their rights under the program's code of conduct; a 24-7 complaint investigation and resolution process, where workers can report violations free from the fear of retaliation; regular and comprehensive farm audits; and legally binding agreements with fourteen of the world's largest buyers of produce that serve to hold growers accountable for violations through swift market consequences for abuse. The elements of the FFP combine, as pillars, to provide a stable structure protecting workers from exploitation and providing an environment of respect in which they can work with dignity.

If a zero-tolerance violation such as forced labor were to be found on a participating farm, that grower would be automatically barred from selling its produce to fourteen of the world's largest buyers, creating a powerful market incentive that, for workers under the FFP's protections, has effectively eliminated the most egregious abuses that have plagued US agriculture for generations. It is this market incentive—or, as supply-chain experts call it, the "power of the purchasing order"—of big buyers like Walmart, Whole Foods, Taco Bell, and McDonald's that truly sets the Fair Food Program apart. Without the knowledge that failure to comply with the program's human rights standards will cost them business, suppliers will always balk at making uncomfortable changes, no matter how necessary. Quite simply, those changes would have been made long ago—and some other industry would have won the ignominious moniker of "ground zero for modern-day slavery"—if that were not the case.

A TRACK RECORD OF SUCCESS

Today, the FFP has established a track record of proven success more than a decade long of remedying, and preventing, farmworker exploitation by empowering workers to speak up against any abuse they may face without fear of retaliation. "The fields have changed," one worker told an auditor who monitors compliance with the FFP's standards by conducting regular audits and complaint investigations as part of the Fair Food Standards Council. "Now we have better wages and better treatment for everyone. Before, there was nothing like that.... We are building a road forward, and we will never go back."

Another farmworker told an auditor: "I am thankful for the program that protects us now. Many years ago, we did not have a voice, rights, or the freedom to raise complaints—but today, with the help of Coalition of Immokalee Workers and Fair Food Standards Council, we can work in peace and with great freedom."

Amalia Mejia Diaz, a former farmworker who came forward to help human rights investigators with a sexual assault case, said, "The work that [the FFP] does makes you feel that you are not so alone in this country. I think many women now have more courage to speak and not remain silent."

The numbers tell the story of the FFP's unique success: more than 3,200 hotline complaints investigated and resolved, nearly ten thousand violations found and corrected, more than 1,200 worker-to-worker education sessions held on the clock on participating farms, and more than $42 million in FFP funds distributed to workers as bonuses on their weekly paychecks, all since the program's inception in 2011. The FFP has forever changed the Florida tomato industry where it was born, and it is extending its impact to more fields, more crops, and more workers every day across the United States and beyond.

Together, a nationwide network of farmworkers and allies has not only rooted out the worst actors in the industry from participating farms— those who for years sexually harassed and beat workers, stole wages, and forced men and women in the fields to work under the threat of violence— but has transformed tens of thousands of farmworkers themselves into the frontline monitors of their own rights. In turn, those workers have provided the fuel for the continued expansion of the program's standards over time to address emerging issues, most recently the COVID-19 pandemic and a heat-stress crisis among workers who labor outdoors.

For example, as climate change makes the summers hotter and thus the work in the fields harder, the program has adapted—through feedback from workers and a unique partnership with growers that evaluates and fine-tunes the FFP through monthly meetings—to include industry-leading heat-stress protocols, mandating shade, water, and rest breaks during the hottest months and giving workers the power to enforce those rights, individually and collectively, if they are violated. Cruz Salucio, a former farmworker and current CIW staff member, says, "Today, unlike when I started in the fields, when there is a great deal of heat and you need to rest, you know you can take action to protect yourself without fear of retaliation; you can use your voice as a worker. Now, as a farmworker, I can leave my bucket for a moment, and know that a short walk away, I will find water, a bathroom, and shade in which to rest. I know my boss will not yell at me to get back to work, or threaten to fire me."

OUTSIDE THE FFP'S PROTECTIONS, ABUSES WORSEN

Meanwhile, on farms outside the Fair Food Program, conditions are deteriorating. With the rise in the use of the H-2A, or "guestworker," visa on farms across the country, exploitative labor contractors are luring farmworkers into insurmountable debt in exchange for access to the temporary work visas. Whatever limited power workers outside the FFP may have had to speak up against this abuse is being stifled. As a result, more and more workers are slowly and silently being entrapped in forced labor.

In 2016, two workers hid in the trunk of a car to escape from a labor camp in Pahokee, Florida, where they were being held against their will, paid abysmally low wages, and threatened with harm to their families at home if they were to complain. As soon as they reached safety, they called the CIW for help. That call launched a multistate forced-labor prosecution, *United States v. Moreno*, that concluded six years later, in December 2022, with the conviction of four labor contractors. Dozens of workers, ensnared by predatory debts charged by the labor contractors, had had their passports confiscated and were made to endure their exploitation under threat of deportation if they spoke up.

Shortly after the convictions of the abusive farm bosses were announced, the Department of Labor revealed that Kroger, a company that has refused to join the Fair Food Program, was a principal buyer of watermelons harvested by workers trapped in the forced-labor operation. In solidarity with those who were plunged into modern-day slavery, hundreds of farmworkers with the CIW and their allies marched fifty miles over the course of five days, from the site of the labor camp in Pahokee to the billionaire enclave of Palm Beach, to call for Kroger, Wendy's, and Publix to join the FFP. The march highlighted the two parallel worlds within agriculture today: the farms operating outside the FFP's gold-standard human rights protections, and those within.

FFP AND WSR EXPAND TO MEET GROWING CHALLENGES

With exploitation increasing outside the program, farmworkers across the globe need the protections offered by the FFP now more than ever. To help guarantee the human rights of all farmworkers, the FFP and other worker-driven social responsibility programs are breaking new ground, both in the United States and abroad. From its roots in the Florida tomato industry, the FFP today covers nine crops in ten states and is beginning to scale internationally. In early 2023, a partnership between the CIW and the Bureau of International Labor Affairs (part of the US Department of Labor) was launched for the expansion of the FFP in Chile, Mexico, and South Africa. At the same time, the CIW is joining forces with the International Transport Workers' Federation and the University of Nottingham's Rights Lab in a groundbreaking collaboration to explore the development of a Fair Fish Program in the UK fishing industry, with fishers, vessel owners, and retailers there to address a growing human rights crisis on the high seas.

Other WSR programs are taking root as well. Here in the United States, dairy workers in Vermont inspired by the FFP launched the Milk with Dignity program and secured Ben & Jerry's as a buyer backing their human rights. Now, dairy workers on scores of farms in Vermont can harness the purchasing power of one of their employers' top customers to monitor and enforce their own rights. Likewise, construction workers in Minnesota have launched another WSR program, dubbed Building

Dignity and Respect, and are actively campaigning for better working conditions on their job sites.

On the international front, in the aftermath of the horrific Rana Plaza collapse in 2013—which killed over one thousand garment workers in Bangladesh—unions, human rights groups, and corporations partnered to form the Accord on Fire and Building Safety in Bangladesh, which is celebrating its first decade in operation and now protects over one million garment workers and counting, expanding into Pakistan this past year under the new banner of the International Safety Accord. Meanwhile, garment workers in Lesotho banded together to take on systemic sexual harassment and assault at work and formed their own WSR program. Day by day, the WSR model is gaining steam across the globe.

What unites all these movements is long-overlooked workers coming together to fight for their basic human rights—and their struggle to obtain the power necessary to overcome generations of widespread exploitation and abuse. Meanwhile, back in the dusty crossroads town of Immokalee, where workers occupied the central square in a general strike that ignited this growing movement thirty years ago, farmworkers still toil in the fields to feed America, but they've also given the world something else: hope. Hope for a better future in the form of a solution to forced labor and other forms of exploitation that are so often obscured by sprawling and opaque corporate supply chains. The human rights revolution sparked by their pioneering work has only just begun to spread to new continents and new industries, but its future is bright, and the potential of its life-changing mechanisms to bring dignity and respect to forgotten workplaces around the globe is all but limitless.

10.

Politics on Your Plate: The Problem and the Promise of US Food Policy

By Cory Booker

Since 2013, Cory Booker has served as a US Senator from New Jersey. He currently sits on the Senate Committee on Agriculture, Nutrition, and Forestry and has served as chair of the Subcommittee on Food and Nutrition, Specialty Crops, Organics, and Research. In those roles, he has spearheaded efforts to empower family farmers, improve the affordability and availability of nutritious foods, and break the grip of massive corporations that are making record profits at the expense of workers, the environment, and our health. Senator Booker believes that our food system is deeply broken and is fighting to invest in practices that make healthier food more available, keep family farmers in business, and rein in the abuses of factory farms and the largest agribusinesses.

Food has always played a central role in my life. One of my most vivid childhood memories is digging up the backyard of my parents' home in the suburbs of New Jersey with my grandfather and planting a garden. To this day, I remember mealtime conversations with high school teammates and coaches, shared after games in dimly lit restaurants where we spoke with youthful exuberance of our future plans and dreams. Even now, I savor the opportunity to break bread with friends and loved ones. Though my journey with food has taken me from a boy who enjoyed the sumptuous turkey his mother cooked on Thanksgiving to a vegan (who still cannot walk idly by an apple crumble!), one thing has remained constant: my love of food.

I love food not only because it tastes good, but also because it possesses a transcendent quality. It's felt in our mouths, our stomachs, and our hearts, whether we're following a recipe passed down through generations that connects us with our ancestors or trying a new cuisine from a land unknown. Food can be expressive, a love language that's communicated when we make a grocery run before cooking for friends and family. It has the ability to define cultures, to link them, and to bring people together. Some of our most sacred religious practices revolve around food, from Jewish seders to iftars during the Islamic holy month of Ramadan to the Holy Communion celebrated by Christians. In short, as the renowned chef James Beard said, "Food is our common ground, a universal experience."

Unfortunately, though we derive an indescribable joy from food, peeling back the curtain at how a meal gets to most Americans' tables reveals that the federal policies that govern the production of food in the United States are failing us. These policies contribute to a savagely broken American food system that is hurting urban and rural communities, destroying the livelihoods of independent family farmers and ranchers, and harming Americans. The only winners from this food system are the massive, consolidated, multinational corporations that dominate our food industry and too often dictate food policy.

ACCESS TO HEALTHY FOODS: THE FOOD DESERT CRISIS

Since 2021, I have served on the Senate agriculture committee, which creates and oversees our national food and farm policies. But I didn't learn about America's food system in the halls of Congress. Instead, my education began in what some might consider an unlikely place: the urban city blocks of Newark, New Jersey, a city that I've called home for more than twenty-five years.

When I became mayor of the city in 2006, my new administration had to address pressing issues such as lowering violent crime, building more affordable housing, and creating greater and fairer economic opportunities for all. But there was another pernicious problem that frustrated residents brought up with me again and again when I met them on sidewalks, at public events, or in town halls. Pulling me aside, they would describe the trials they endured to purchase healthy, affordable food for themselves and their families. Some had to commute nearly five miles to shop, while others spent their hard-earned wages on public transportation to travel to a grocery store outside of the city. More than a decade had passed since a new supermarket had opened in Newark, and large sections of the city were "food deserts"—places with limited access to affordable, nutritious food.

Through a combination of grant funding and investments, we enacted change, and in 2012, a new supermarket finally opened its doors. Then, a year later, a new ShopRite grocery store broke ground, quickly followed by a Whole Foods, and local bodegas were provided with support to offer fresh, healthy options beyond the snacks that people already loved. Through the reforms we made, the city's food system improved, and while there was still much work to be done, residents were more easily able to afford and obtain nutritious food.

I am proud of the local food initiatives that my administration championed. But I always think of how much more we could have achieved if we'd had the weight of the federal government behind us. We often had to push against the inertia of federal policies that hindered rather than helped our cause—policies that perpetuated the problems that plagued us more often than they solved them. It became obvious to me that solely enacting solutions at the local level would not yield the wholesale, comprehensive changes we sought. To change our nation's food system, our federal food and farm policies have to change too.

As of 2021, an estimated nineteen million Americans, disproportionately people with low incomes and people of color, have limited access to a supermarket or grocery store.[1] Food deserts aren't an exclusively urban phenomenon; they exist in countless rural places too. In many of these communities, the absence of grocery stores has been filled by fast-food restaurants that are propped up by federal subsidies that make their ingredients artificially cheap. Although federal dietary guidelines tell Americans that more than half of their diet should consist of fruits and vegetables, only a small fraction of our federal farm subsidies actually goes to farmers growing these healthy foods. Billions of taxpayer dollars instead go toward subsidizing commodity crops like corn and soybeans, which often end up in fast foods and other ultra-processed products that are making Americans sick. The University of Southern California's Bedrosian Center succinctly summarized the allocation of federal subsidies as one that "drives down the prices of fast food items, leaving fresh produce more expensive and less accessible to low-income areas."[2]

Clearly, there is a misalignment between the foods our government chooses to subsidize and what Americans are encouraged to eat as part of a healthy diet. This is no accident: agribusinesses spent $750 million on national political candidates from 2000 to 2020 and $2.5 billion on lobbying between 2000 and 2019.[3] These powerful special interest groups have fought to ensure that federal subsidies continue to flow to commodity crops and the largest farms.

TOO BIG, TOO POWERFUL, TOO UNACCOUNTABLE

The lack of access to healthy foods can also be traced to another federal policy failure: the federal government's complicity with the increased corporate consolidation in our food system. This consolidation has led to fewer consumer choices and a handful of big multinational companies with massive economic and political power now having undue influence over federal food and farm policy. An example that illustrates this problem is the corporate consolidation that's happened among grocery store chains.

Consolidation in the grocery store industry can be traced to the 1990s, when retail giants such as Walmart and Target entered the scene. During this time, mergers and acquisitions accelerated as large, wealthy corporations consolidated and increasingly behaved like monopolies—a trend that continued well into the 2000s. In the span of just over two decades, the national market share of the four leading retailers in the industry rose from 23 percent in 1993 to 55 percent in 2014.[4]

None of this would have been possible if not for the federal government straying away from enforcing antitrust laws that were already on the books. The new laissez-faire approach was inspired by the legal philosophy of Robert Bork, former solicitor general of the United States during the Nixon administration and judge on the powerful US Court of Appeals for the DC Circuit, who argued that antitrust enforcement should be solely concerned with promoting "consumer welfare." In other words, Bork contended that market concentration was acceptable so long as it would lead to greater efficiency and lower prices. His thinking, which influenced subsequent Republican and Democratic presidential administrations, largely ignored the deleterious effects of corporate consolidation on small businesses, workers, and communities.

As retail giants entered and solidified their presence in the grocery store industry, smaller mom-and-pop grocers were powerless to compete and ultimately forced to shut their doors. When supermarket chains merged, as Albertsons and Safeway did in 2015, store closures followed as two previously rival companies consolidated operations. In the process, workers lost their jobs and communities were left with fewer options for purchasing affordable groceries.

Consolidation in the retail store industry is emblematic of the consolidation that is alarmingly occurring across our entire food and farm system. Over the past four decades, the top four largest pork packers have seized control of 70 percent of the market, up from 33 percent; in the same time period, the top four beef packers have expanded their market share from 25 percent to 85 percent.[5] Their ever-growing influence has been especially devastating to rural communities where these large, multinational corporations have externalized their costs and privatized their profits, a phenomenon I witnessed firsthand when I traveled to Duplin County, North Carolina, a few years after my election to the US Senate.

In Duplin County, I met with Black residents whose lives were affected every day by concentrated animal feeding operations (CAFOs).

These look nothing like the bucolic image people have when they think of a farm; instead, they are best described as huge animal factories. In eastern North Carolina, I saw countless CAFOs, metal sheds no taller than a single-story house that stretched the length of a football field. Inside, thousands of hogs were confined in inhumane conditions; outside, one could smell the stench of hog waste long before seeing the massive cesspools in which this waste was being collected, manure "lagoons" larger than city blocks. CAFO operators would regularly spray the liquid waste from the lagoons onto adjoining fields. Nearby residents described to me the litany of problems this would cause for them: how the waste would leach into groundwater or drift into their backyards, forcing people to shelter in their homes and preventing them from gardening, enjoying a cookout, or even hanging their clothes out to dry.

Though the first CAFOs came into existence in the mid-1900s, long before Bork formulated his legal philosophy on antitrust enforcement, their proliferation and continued presence can again be traced to the failure in recent decades of our federal government to enforce antitrust laws. As corporate agribusinesses have grown in size and power, they have pushed for production models that harm the environment and public health but fit neatly into their industrialized supply chain. This has forced farmers to get big or get out and deprioritized stewardship of natural resources and communities. North Carolina alone has more than 2,100 CAFOs, many located in Duplin County and nearby Sampson County.[6] Most CAFOs are built in low-income areas where land is cheaper and where nearby residents have less political power to protest their harmful operations. Studies show that African Americans, Latinos, and Native Americans are more likely to be living within three miles of a hog facility than non-Hispanic whites. Additional research shows that areas of high poverty and areas where high percentages of non-white people live have seven times more industrial swine facilities.[7]

It's also becoming increasingly clear that Bork's consumer welfare standard—the theory that marketplace concentration would lead to greater efficiency and lower prices, and is thus acceptable—is not playing out as he envisioned. Even with lower production costs, megacorporations fail to pass on benefits to American consumers, instead choosing to raise prices and deliver profits to investors. Since 2018, prices of food and groceries have outpaced overall inflation, and food prices increased faster in 2022 than in any year since 1979. Unfortunately, federal courts are still heavily influenced by the consumer welfare standard and have been reluctant to issue rulings challenging the doctrine, despite increasing evidence of the harm it causes.

When looked at from the perspective of consumers, independent family farmers and ranchers, or vulnerable communities, the hyperconsolidation of our food system has been damaging. The promised benefits of consolidation—namely, lower prices for consumers—have failed to pan out.

A FOOD-DRIVEN HEALTH CRISIS

As large corporations have taken control of almost every aspect of our food system, the United States has experienced a nutrition crisis that is wreaking havoc on the lives of Americans. Recent decades have seen skyrocketing rates of diet-related diseases such as type 2 diabetes, heart disease, and certain cancers; half of the current US population is prediabetic or has type 2 diabetes. The numbers are even more grim for Black people, who are at a 77 percent higher risk of acquiring diabetes than their white counterparts and are twice as likely to die from the disease.[8] As a result, nearly one out of every four dollars of our federal budget now goes toward health-care spending.[9] With young people experiencing higher rates of diabetes and obesity as well, that cost is only set to rise and threaten the long-term well-being of our nation.[10]

The heightened prevalence of diet-related diseases is in no way reflective of any moral shortcoming of individuals. Instead, it is driven by corporate greed and a lack of government oversight to monitor large, consolidated, multinational companies. The failure to prevent mergers and acquisitions in the retail grocery industry has meant the closure of supermarkets in underserved urban and rural communities, creating food deserts and making it hard for people to find, let alone afford, healthy, nutritious food. In lieu of subsidizing fruits and vegetables, our government subsidizes commodity crops that ultimately end up as high-fructose corn syrup in nutrient-poor and addictive ultra-processed foods. Approximately $2 billion per year is spent by the food-and-beverage industry in marketing these foods, making them seem irresistible to the point that 60 percent of the calories in the food supply are now ultra-processed.[11]

One way that our government can begin to address the growing nutrition crisis is through the US Dietary Guidelines for Americans, established more than four decades ago as a joint effort between the USDA and the Department of Health and Human Services (HHS). Updated every five years, most recently in 2020, these public guidelines define how several federal food, nutrition, and health programs in our nation are implemented. Making sure that the guidelines are based on the most sound and up-to-date science would provide Americans with accurate advice on what constitutes a healthy diet, what foods we should eat more of, and what foods we should avoid.

It turns out, however, that the process of developing our federal dietary guidelines often strays away from sound science, again influenced by corporate interests. After scientists expressed concerns over how the guidelines were conceived, and following a congressional hearing questioning the scientific integrity of the process, Congress mandated that the National Academies of Sciences, Engineering, and Medicine (NASEM) examine the issue. As part of its 2017 report, NASEM recommended that the USDA and HHS strengthen their scientific methodologies and periodically update them to align with best practices, change the selection process for the expert panel that informs the guidelines, and justify any differences between adopted guidelines and those recommended by the expert panel.[12]

The NASEM report also recommended that additional guidelines be formulated specifically for Americans with diet-related diseases. When first commissioned in 1980 by Congress, the dietary guidelines were required only to address the "general public," which at the time was relatively healthy compared to today; as a result, the earliest dietary guidelines focused on preventing the onset of diet-related diseases. Forty years later, with diet-related diseases now affecting the majority of the US adult population, the guidelines should theoretically be focused on treatment and management of these illnesses. Yet USDA officials have maintained that this purpose is beyond the scope of the dietary guidelines, which raises questions over whether current and future guidelines are truly applicable to America's population. As for the other recommendations NASEM suggested, only a fraction have been adopted.

SOLUTIONS WITHIN OUR REACH

Up until now, this chapter has traced a few key factors that contribute to our nation's broken food system and nutrition crisis, from the misalignment of our federal farm subsidies, to corporate consolidation and a lack of antitrust enforcement, to the systemic issues that affect the creation of America's dietary guidelines. Just as the federal government has been part of the problem in creating our broken food system, it can—and must—be part of the solution to fix it. I'm especially encouraged by new initiatives that are gaining momentum and that I believe have the potential to set our country on a course for change.

One such initiative was the White House Conference on Hunger, Nutrition, and Health hosted by President Joe Biden in September 2022. This was the second conference of its kind, modeled after one held more than fifty years ago that was geared toward addressing widespread hunger and severe malnutrition in impoverished communities across America. Out of that first conference came landmark programs such as the National School Lunch Program, the School Breakfast Program, and the Special Supplemental Nutrition Program for Women, Infants, and Children, which were life-changing for millions of struggling Americans and significantly reduced hunger and malnutrition. Yet, despite the historic progress made more than fifty years ago, hunger still persists in America, and we now face an alarming crisis of nutrition insecurity, as millions of Americans are unable to access healthy, safe, and affordable foods.

The 2022 conference sought to address these dual crises by emulating the bold, whole-of-government approach taken by the first one. I remember the enthusiasm in the room as scientists, public health experts, and government officials convened to discuss areas of collaboration and policy solutions. On the same day as the conference, the Biden administration released a new national strategy to end hunger in America, increase healthy eating, and promote physical activity to significantly reduce diet-related diseases by 2030. Included in the national strategy were pilot programs that make medically tailored meals covered as benefits in Medicaid and Medicare for people with diet-related diseases. These programs took inspiration from an idea known as "food as medicine," which is the understanding that the food we eat is intrinsically tied to our health and an essential element of preventing and treating diet-related diseases. If successful, the programs can eventually lead the way to broader coverage of medically tailored meals, nutrition coaching, and healthy produce prescriptions in our health-care system, leading to significant health benefits for participants as well as lower health-care spending.

The White House conference represents just a part of the policy change we need. A few weeks before attending the conference, I completed a personal challenge: from Independence Day to Labor Day, I decided to stop consuming foods that contained any added sugars. After a Fourth of July meal that concluded with a slice of apple pie, I began my journey, sharing my progress on social media and encouraging others to join. My pitch detailed how the production of ultra-processed foods in the United States illustrates many of the problems mentioned earlier in this chapter. Sweeteners such as high-fructose corn syrup are heavily subsidized by our government, lowering their cost and making them a common ingredient in items throughout our grocery stores. Partly as a result, a typical American consumes about seventeen teaspoons of added sugars and sweeteners per day, roughly 50 percent more than is recommended by the World Health Organization.[13] And by a wide margin, the United States ranks first in sugar consumption compared to other countries, contributing to increased incidences of cardiovascular disease, diabetes, dental decay, and other diet-related illnesses.

When I began the challenge, I found that it wasn't so much the lingering desire for sugar that made it difficult (though the first few weeks were certainly hard) but the task of finding foods that didn't contain added sweeteners. I remember scrutinizing grocery items and turning them over to read their ingredient list before returning them to the shelf in frustration. Even those grocery items that one might expect to be devoid of added sweeteners—such as hot dog buns or purportedly "sugar free" barbecue sauce—contained sugar in the form of high-fructose corn syrup or sucrose.

Simply stated, it shouldn't be this difficult for consumers to make healthy choices when going to a supermarket. For someone worried about hypertension, or a person with diabetes concerned about their sugar intake, clear labels on the front of grocery store items would provide them with the necessary information to make more informed decisions. Days after I completed my challenge, the Biden administration, as part of its national strategy, announced steps at the White House conference to make this a reality by calling on the US Food and Drug Administration (FDA) to use its existing legal authority to implement front-of-package food labeling to warn shoppers of high levels of added sugar and salt.

The burden now falls on lawmakers and policy officials to work with the Biden administration to build on the provisions of the national strategy. Hopefully, when we reflect back years from now, the legacy of the second White House conference will rival the transformative policies that came out of the first conference hosted in 1969.

In the same way that we looked to history as a guide for the recent White House conference, we can look to history to inspire us to address the rampant corporate consolidation that has gripped our food system just as it did in the early 1900s. Back then, the Federal Trade Commission issued a report detailing the manipulative practices of the "Big Five," the five largest meatpackers of the time.[14] According to the report, the Big Five had attained such a dominant position "that they control at will the market in which they buy their supplies, the market in which they sell their products, and hold the fortunes of their competitors in their hands." Just two years later, Gilbert Haugen, an Iowa congressman who was chair of the US House Agriculture Committee, introduced the landmark Packers and Stockyards Act, legislation that took direct aim at the giants of the meatpacking industry. As stated by Congress, the act sought to promote fair competition and trade practices, safeguard farmers and ranchers, and protect consumers from monopolistic practices.[15] Today, more than one hundred years after the Packers and Stockyards Act was first implemented, the law has not been meaningfully updated by Congress to address a completely changed marketplace. Coupled with the government's departure from its mandate to enforce antitrust laws, this has rendered the legislation largely ineffective for our time. For a measure of how toothless the law has become, consider that the meatpacking industry today is actually *more* concentrated than it was in 1921.

Fortunately, there is a growing consensus among policymakers that the Packers and Stockyards Act needs reform. The Biden administration has proposed several rules to substantially strengthen and modernize implementation of the law. In the Senate, I've introduced legislation that would go even further. My Food and Agribusiness Merger Moratorium and Antitrust Review Act would put an end to the monopolistic practices of multinational meatpacking companies by placing a moratorium on any new big mergers in the agriculture sector until Congress takes action to update our antitrust laws.[16] The Farm System Reform Act, another bill that I've sponsored, would seek to level the playing field for independent family farmers by placing an immediate moratorium on the construction of new large CAFOs, phasing out the largest ones completely by 2040, and further strengthening the Packers and Stockyards Act to better combat monopolistic practices.[17]

It's important not only to divest power from large businesses but to transfer it back to family farmers who have been devastated by the corporate consolidation of the past decades. In 1950, the farmer's share of every retail dollar was forty-one cents; today it stands at less than fifteen cents.[18] Net farm income for US farmers has fallen by more than half, forcing many to declare bankruptcy or rely on multiple jobs. I've sat around kitchen tables talking to farmers and ranchers who have shared stories of the turmoil and stress they and their families routinely endure. The decline of family farms has also resulted in the demise of rural communities, as stores lose customers, schools become underused and eventually close, career opportunities for young people dry up, and inequalities of wealth and income grow wider.

There already exist small-scale federal government programs that, if expanded and properly utilized, have the potential to reverse these harmful trends, empower small family farmers, and create resilient local and regional food systems that benefit communities. Every five years, the federal government revisits many of these policies when it updates a law commonly referred to as the Farm Bill, which lays out the agricultural and food policy of the federal government. The legislation touches on everything from rural development, trade, conservation, and agricultural research to, yes, food and nutrition programs.

The current iteration of the Farm Bill, passed in 2018, contains a little-known program called the GusNIP (Nutrition Incentive Program), named after Gus Schumacher, a leading figure of the food justice movement. Under this initiative, some participants in the Supplemental Nutrition Assistance Program (SNAP) are able to double a portion of their benefits to purchase fruits and vegetables, which in turn incentivizes farmers to grow healthy food. I've personally witnessed the incredible and holistic impact that the program can have. During my days as Newark mayor, my administration partnered with a local nonprofit to open the Hawthorne Avenue Farm in our city's South Ward. Ten years later, this three-acre space continues to pay dividends for residents precisely because of the GusNIP program. I recently met two women there who told me they used their GusNIP benefits to purchase fresh, nutritious food from the farm, which they incorporated into their diet. They praised the impact of the program on their lives, telling me that the chronic health problems they struggled with were alleviated by their healthier diet, attesting to the promise the program holds for others.

Despite its success in Newark and the potential it holds, the GusNIP program is currently small and underfunded, limiting its ability to reach more places. But what if that were to change in the next iteration of the Farm Bill? Consider this: If we were to allocate billions of dollars into the program—even a fraction of the subsidies we currently provide large corporations to grow corn and soy—more small farmers would have a more sizable market to sell fruits and vegetables, incentivizing them to grow healthy food. At the same time, we would take a step forward to address the misalignment that plagues our federal Farm Bill subsidies by finally reallocating funds to the foods Americans are encouraged to eat and should be consuming instead of those that make them sicker.

CHANGE IS COMING

As the saying goes, we are what we eat. And right now, it's painful to say that what we are eating is making us sicker at every turn; is contributing to gross environmental, health-care, and economic injustices; and is an affront to the values we cherish as Americans. Family farmers are struggling and too often losing their farms, big agriculture conglomerates are getting bigger, and healthy, nutritious food is hard to find and even harder to afford for families in rural and urban communities alike. More than thirty-five million Americans are food insecure, and diet-related diseases are skyrocketing, a consequence of malignant corporate practices and misguided policy decisions.

Yet, despite all this, I am actually more hopeful now than ever before. This isn't a foolish optimism but a firm belief that we possess solutions to these myriad problems. Of course, reforms to the status quo are being challenged by powerful interests working every day to stifle change and perpetuate a system that benefits only them. But from my conversations with the people at the heart of our food system—whether they be farmers, animal welfare activists, food service workers, environmental justice advocates, health-care professionals, or anyone who loves food—I increasingly feel that we are approaching a tipping point. There's a growing recognition of the injustices plaguing our food system and the urgent need to achieve wholesale reform.

The truth is, change doesn't come from Washington, it comes *to* Washington. Right now, voices are rising from the bodegas of Newark, the fields of North Carolina, the farms of the Midwest, the ranches of the West, and everywhere in between. In the halls of Congress, you can hear the growing chorus, demanding change. It's time for us to listen and act.

11.

The Future of Food Is Blue

By David E. Kelley and Andrew Zimmern

Multi-award-winning writer and producer David E. Kelley is behind some of America's most groundbreaking and distinctive television programs, which address contemporary issues such as social justice, diversity, and privilege. He is the creator of the Emmy, Peabody, and Golden Globe Award–winning shows *Boston Legal*, *The Practice*, and *Ally McBeal*, plus the critically acclaimed dramatic series *Boston Public*, *Chicago Hope*, and *Picket Fences*. Kelley's adaptation of Liane Moriarty's book *Big Little Lies* received multiple awards, including an Emmy for Outstanding Limited Series as well as four Golden Globes. His latest projects include HBO's most-watched series of 2020, *The Undoing*, Hulu's most-watched original series, *Nine Perfect Strangers*, and the series *Love & Death* for HBO Max. In addition to his incredible TV career, Kelley is the founder of Riverence, the largest grower of steelhead and rainbow trout in North America, with twenty sustainably managed farms in Idaho and Washington State.

An Emmy-winning and four-time James Beard Award–winning TV personality, Andrew Zimmern is also a chef, a writer, and a social justice advocate. As the creator, executive producer, and host of the *Bizarre Foods* franchise, *Andrew Zimmern's Driven by Food*, and the Emmy-winning *The Zimmern List*, he has devoted his life to exploring and promoting cultural acceptance, tolerance, and understanding through food. In 2020, Zimmern moved the needle with his award-winning limited series on MSNBC *What's Eating America*, and, in 2021, *Family Dinner* premiered on Magnolia Network, earning Emmy nominations in 2023 for best show and best host. In 2023, he is judging the epic culinary battle *Iron Chef: Quest for an Iron Legend* on Netflix and teaching us how to cook sustainably from the wild with Outdoor Channel's *Andrew Zimmern's Wild Game Kitchen*. Zimmern is a sought-after speaker on the topic of restorative water farming. In his travels, he has seen firsthand the plight of our oceans, its adverse impact on local communities, and the potential of restorative water farming to boost local economies, regenerate marine habitats, and help feed a growing world population.

What's the common thread that connects a storyteller and a chef? They both tell stories, through words on a page or through ingredients on a plate. They both touch souls, provide bonding and enriching experiences. And, in the context of our food system, they both play vital social and cultural roles to educate and open our minds.

Passionate about feeding people with hearty foods and stories, the two of us—one a storyteller, avid fisherman, and trout farmer, one a chef turned writer, TV personality, and storyteller—have plenty of food for thought on the future of food systems and the crucial role that "blue foods"—foods from the world's oceans, lakes, and rivers—can play in it.

Since 1986, David E. Kelley has incorporated topics of cultural and environmental impact into his writing, using his platform to talk about important issues like climate change in ways that are understandable and compelling for everyone. He has been inspired by the efforts of other filmmakers to call attention to the urgent environmental issues of our time. For example, Ali Tabrizi's groundbreaking documentary *Seaspiracy* brought to light specific fishing and aquaculture practices that need to end and highlighted the fact that we urgently need to find new ways to protect the oceans, which serve as Earth's life-support system.

But *Seaspiracy* also told a story that is incomplete, because there are already a range of impactful solutions at work in the water. In reality, something as intricate and interconnected as the oceans' ecosystem cannot be viewed as one problem and one solution. David is determined to use his gifts as a filmmaker and activist to bring the complex realities of this challenge to millions of people, thereby helping to realize the day when real solutions are taking shape around the world.

Andrew Zimmern cooks food for living, an adventurous process that starts way before he gets to the kitchen. The workers who spend time in the fields producing the ingredients—as well as the countless others who harvest, process, package, and distribute them—all work behind the scenes, part of the process that is usually invisible to the consumer but with results that are more than tangible. Finding ways to empower these workers while also feeding the world and protecting the environment that makes it all possible is the driving cause that brought Andrew to this conversation.

Andrew has been a longtime advocate of cooking seafood. His first job was in 1975 at the Quiet Clam in Long Island, shucking oysters and cherrystones. He grew to understand the health benefits of eating more seafood. In his role as a chef, he is optimistic about the transformative part that the culinary industry can play today in sharing all we know about the importance of seafood for our own health and that of the planet.

In 2021, the two of us teamed up, embarking on a journey in search of new sources of sustainable sustenance for the world's hungry billions. We called our project Fed by Blue. Its goal is to help trigger a blue revolution built on the scientifically proven knowledge that sustainably sourced aquatic foods can change the future of food systems by providing quality nourishment, lowering world hunger, and having a positive impact on our oceans and planet.

WHAT ARE BLUE FOODS?
HINT: NOT BLUEBERRIES!

Around the globe, more than three thousand species of aquatic animals and plants are used for human nutrition. Called blue foods, they are produced through a wide variety of systems: from streams, rivers, lakes, wetlands, seas, and oceans to land-based, innovative aquariums. From these abundant waters, we catch and cultivate many different aquatic animals, plants, and algae. Diverse and nutritious, blue foods are already a cornerstone of the global food system, providing vital sources of nutrition and protein for more than 3.3 billion people worldwide and livelihoods for hundreds of millions.[1] Humankind will continue to rely on them as our population grows.

But they have even greater potential, provided we manage and nurture them responsibly. In the words of the UN Sustainable Development Goals, "The global action agenda for the food system should embrace the priorities for a sustainable blue food system including the need to end overfishing and the harmful subsidies that drive it, to preserve and restore marine and coastal ecosystems, and to ensure that small-scale fishers have access to resources and markets."

When responsibly caught or harvested, blue foods provide an abundance of restorative opportunities for environmental, economic, and global health. Responsibly produced blue foods mean improved wild fisheries management and fishing practices. That requires innovation, cleaner practices, and using fewer resources to grow and harvest blue foods. It also includes fair thinking by maintaining and creating decent livelihoods for our water fishers, farmers, and harvesters.

REIMAGINING THE FUTURE OF OUR FOOD

As citizens of this planet, we are living in a time of endangered oceans, climate crisis, world hunger, failing aquaculture, and biodiversity loss. Yet, at the same time, we are witnessing new, purposeful waves of desire to reimagine food systems, production, ingredients, and their social and environmental impact. People are more and more open to learning how to manage our waters in a way that's better for communities, wildlife, and the environment.

Through encouraging restorative practices, not only will we relieve pressure on nature, but we can rebuild our relationship with blue foods while meeting the nutritional needs of a growing world population. Along with many others, the two of us are dedicated to the vast ocean of opportunities, solutions, and ideas that already exist and to creating ripples of change wherever we can.

Luckily, we are already witnessing remarkable efforts underway. Led by various thought leaders—entrepreneurs, farmers, environmentalists, technologists, chefs, nutritionists, and storytellers—innovative and groundbreaking solutions are being explored that can return the world's waters to abundance while also supplying our population with valuable food resources. Significant advances in practices and technology have made it possible to responsibly manage the blue foods sector, creating real potential to restore environmental health and improve human well-being.

Rethinking how and where we grow our food is critical for our climate future. There are plenty of solutions, including no-take zones in the ocean, regenerative water farming through integrating bivalves and seaweed, transitioning from traditional agriculture to responsible aquaculture, improved water-monitoring technologies, and more efficient feeds for fish farming. Together, let's dive deeper into some of those solutions.

RETURNING FISHERIES TO ABUNDANCE: THE IMPORTANCE OF NO-TAKE ZONES

Through working on our nonprofit initiative Fed by Blue and filming our television series for PBS, *Hope in the Water*, we've seen how some endangered coastal and marine zones around the world need immediate, urgent, and strict protection. Just as we have national parks and preserved areas inland, there is an acute need to protect threatened water areas.

Marine protected areas are now being created exactly with that aim in mind—to designate marine spaces where wildlife, natural habitats, and water quality are highly protected. Inside a marine protected area, limits on certain damaging activities are rigorously applied, from oil drilling to discharging wastewater and harmful fishing practices.

So far, however, the scope of marine protected areas is insufficient. While more than 15 percent of the world's land area has some form of management or protection, only 4.8 percent of the global ocean is included in existing marine protected areas. Expanding the zones functioning as marine sanctuaries will be a powerful tool in rebuilding and sustaining fisheries and revitalizing the surrounding ecosystems back to abundance.[2]

For example, scientists have discovered that marine protected areas can help globally in combating the effects of climate change. A study published in *Nature* in March 2021 found that bottom trawling—a widespread fishing practice that involves dragging weighted nets across the sea floor—releases annually as much carbon dioxide as the entire aviation industry. The practice is also controversial because the nets can sweep up juvenile fish and many other ocean species. Sole, cod, crab, and scallop are all caught using bottom trawling.[3]

Such destructive fishing practices are banned in most marine protected areas. These sanctuaries allow species to thrive, grow, and reproduce, encouraging the restoration of healthy marine populations both inside and beyond the protected areas themselves.

In addition to marine protected areas, there are more restricted types of ocean conservation areas named no-take zones, set aside by governments, where the key intention is to help restore fisheries and habitats. No extractive practices are allowed in these areas. That includes any action that removes any marine resource, covering everything from fishing to the hunting of marine mammals to mining, oil drilling, and even shell collecting. There are many instruments that can be used to establish these limits, from laws and regulations to voluntary agreements and codes of conduct.

No-take zones interact with natural processes to regenerate marine life in powerful ways. In contrast to humans, who become less fertile as they grow older, female fish produce more eggs as they age. What's more, fish spawned in no-take zones from more mature female fish have a greater chance of survival and higher growth rates. Those two factors can support replenishing fisheries outside of no-take zones in two ways. First, when healthy larvae are carried by currents to fished areas, they can replenish populations accessible to fishers. Second, if there is overcrowding of fish in a no-take zone, they spill over into surrounding areas. As a result, the surrounding fisheries can produce up to three times more fish.[4]

Initiatives across the globe now support constant, long-term efforts toward establishing more marine protected areas and no-take zones. One such campaign, spearheaded by Howard Wood and Don MacNeish, has given citizens a voice in a debate long dominated by the commercial fishing industry.

In 1984, the Scottish government allowed bottom trawlers to their shores. Not surprisingly, the marine ecosystem around the Isle of Arran—once known for plentiful fishing of herring, cod, haddock, and turbot—steadily collapsed as destructive dredgers and trawlers clear-cut the seabed floor. In addition to damaging the seafloor, these activities destroyed coral and kelp forests—vital zones that provide a lot of nourishment to local fish and shellfish.

Wood and MacNeish observed the impact on the aquatic habitat. "When I first started diving," Wood says, "on most dives we would see an abundance of rays, flat fish, and turbot. But within ten to fifteen years, we were lucky if we would see one every few months. Eventually, we wouldn't see any at all."

Wood and MacNeish launched a lobbying campaign with the goal of establishing the first no-take zone in Scotland. MacNeish had earlier witnessed the incredible results of a no-take zone implemented in New Zealand. After thirteen years of fighting to win over the community, lobbying the Scottish government, meeting with scientists, and finding common ground with area fishers' associations, the first no-take zone in Scotland was finally implemented in 2008. In collaboration with the Community of Arran Seabed Trust, Wood identified Lamlash Bay as a key habitat for regenerating marine wildlife. To enforce the new rules for protection of the area, Wood enlisted local divers to work with academic scientists in monitoring the coastline. The effort yielded dramatic recovery of seaweed beds, corals, and juvenile scallops in just a few years. Quicker than expected, impressive revival was noticed in species of mollusks and finfish, returning the area to abundance with more fish and seafood for the fishermen to harvest outside of the no-take zone. Impressed with the results, the Scottish government announced thirty new marine protected areas in the summer of 2014.[5]

With such proven outcomes from marine protected areas and no-take zones, an international effort has been launched to protect 30 percent of our oceans by 2030. Everyone can join in to support this effort, since the oceans belong to all of us.

A study in *Nature* has identified a global network of areas that, if protected by 2030, would safeguard over 80 percent of the habitats for endangered marine species. Raising the level of protection around the world would have a dramatic effect on biodiversity, fish populations, and carbon dioxide levels.

RETHINKING THE NORM: REVOLUTIONIZING THE FISHING INDUSTRY THROUGH REGENERATIVE WATER FARMING

Sometimes, it's too late to protect marine areas where drastic changes have already impacted the local habitat. Instead, efforts toward smart adaptation and mitigation are required. That's what happened in Maine.

For hundreds of years, coastal Maine has been renowned for one major local resource: the lobster. For these communities, lobstering is deeply woven into every aspect of life; tax revenues, jobs, and the state's identity depend on it. But climate change has caused Maine's coastal waters to warm, and, as a result, the underwater life and the economy built around it have shifted dramatically.

Over the course of the last thirty years, generational lobster fishermen have seen a dramatic change in their fishery. The Gulf of Maine is warming at a rate of 0.09 degrees Fahrenheit per year, faster than 96 percent of the world's oceans. The increasingly warm temperatures have pushed the lobsters to migrate north in search of much needed colder and oxygen-rich waters. The impact has been profound. This migration has changed the season for the lobster fishermen from year-round to only six months of the year, leaving families without jobs and income.

Organizations are stepping up with the mission to support the livelihood of coastal communities. One of them is the Island Institute, which helps local communities thrive by educating lobstermen on sustainable aquaculture. As a result, there are now many options for off-season work, including oyster farming, mussel farming, and seaweed farming. All these solutions are based on choosing vital marine species that have a bountiful impact on marine habitats through water filtration, nutrient enrichment, and carbon dioxide absorption.

The burgeoning seaweed aquaculture solution, for example, has been supported in Maine by numerous companies looking to buy the seaweed from these farms and include them in their products.

Atlantic Sea Farms is one such company. It works with twenty-seven partner farmers, and its 2022 harvest brought in just under one million pounds of seaweed. The company's products are sold in more than two thousand stores across the United States, as well as in restaurants and college cafeterias. In 2021, the company was responsible for 85 percent of the line-produced seaweed in the country.

Bri Warner, CEO of Atlantic Sea Farms, explains the importance of their work. "In communities where lobster is everything, how do we prepare for the future along the Maine coast and diversify to face climate change?" she asks. "When you're self-employed and your entire community is dependent on one industry, and you're totally at the whim of Mother Nature, overdependence on one monoculture is very scary."

Warner calls seaweed "a shock absorber against the volatility of the lobster industry." What's more, it is farmed without land, pesticides, or fresh water. "The best thing about kelp," she says, "is that it is the most climate-friendly food you can eat!"[6]

New scientific studies are supporting the climate benefits of aquaculture and highlighting the promising potential of seaweed growing and farming. For example, intercropping seaweed with some types of mollusks—an Indigenous practice inherited from centuries ago—has the potential to reduce carbon emissions created during shellfish farming. Even more encouraging are some of the new developments with the use of seaweed in end products that have climate-positive correlations, such as incorporating seaweed into cow feed to reduce methane emissions from cow "burps" or using seaweed as a raw material in bioplastics and biochar.

Climate scientists and researchers are also discovering the ability of seaweed aquaculture to mitigate the local effects of ocean acidification. One of the most significant impacts of climate change on ocean and coastal ecosystems, ocean acidification happens when too much carbon dioxide (CO_2) is absorbed in the seawater, which leads to changes in its chemical balance. Like plants on land, seaweed uses CO_2 for photosynthesis, which means that as it grows, it removes CO_2 from the surrounding water and thus reduces its acidity. This new knowledge has been confirmed by studies in the United States, Chile, and China, where seaweed farms provide a halo effect to surrounding water by lowering the acidity levels.

When talking about seaweed, we need to highlight kelp, a nourishing sea vegetable that is grabbing the attention of chefs and food producers globally. Many regions around the world are showing interest in getting involved in kelp growing.

In the state of Alaska, kelp farming totaled $1.8 million in sales in 2020 and is forecast to be a $100 million industry by 2040. The industry is already spawning entrepreneurs like Lia Heifetz. Cofounded by Heifetz in 2016, Barnacle Foods is using kelp to make products such as salsa and hot sauce. "We were really motivated to grow a business that could provide a market for kelp farmers and harvesters," Heifetz said. "It's not a cultural norm to have seaweed on our plates in America, but there would be a big upside if that were to happen here." Her line of salsas, rubs, and pickles are natural—and delicious—alternatives to the conventional versions of these products.

On the island of Savai'i in the South Pacific, Sapeti Tiitii is also working hard on cultivating seaweed as part of her mission to help the island communities transition toward a more sustainable local economy—a much needed adaptation after their fisheries have been fished out. Enjoyed throughout generations, local sea grapes, called *limu*, are a favorite delicacy that grow in shallow water and clear, reef environments. In addition to their tasty flavor, they are rich in iron and potassium and have multiple health benefits.

Growing sea grapes is easy to scale and has good income-generating potential, which is why Tiitii goes from village to village discussing the possibilities that seaweed can bring to the region. To convince her fellow islanders, she shows them packages of seaweed snacks from countries like the United States and Australia and tells them how much they can earn (about $3.90 to $5.80) for three pounds of *limu* seaweed. Tiitii is also introducing a method of farming that submerges and suspends trays of *limu* seeds between stakes off the shore. It takes four to six weeks to grow, and, while it grows, the area around the *limu* farms transforms because the fish return.[7]

These inspiring examples from Maine, Alaska, and Savai'i are not isolated cases. Economic aid organizations are recognizing seaweed as an internationally promising industry for a multitude of uses. The islands of Fiji, Samoa, and Kiribati are already proving that seaweed is an aquatic multiuse product, and growing it is helping to support public health, women's empowerment, and environmental sustainability.

In the United States, it is estimated that seaweed production will surpass potato production by 2040. In Chile, seaweed farming has increased by over 400 percent in the last ten years, and dehydrated seaweed products are now the country's second most exported commodity, with more than two-thirds of those exports going to China and the remainder to Canada, Denmark, France, Japan, and Norway.

Coastal communities are the first to be harmed by the impact of climate change on weather, coasts, and oceans. Seaweed is not only a new source of income for them but a mighty crop that can protect coastlines from weather events, provide a boost for local economies, and offer a nutritious source of food in the face of climate change.[8]

MODERN AQUACULTURE AND THE STORY OF RIVERENCE TROUT

As we witness the loss of fisheries, another practice that can stimulate the return to blue foods abundance while also feeding and supporting local communities is fish farming or aquaculture—the rearing of fish, usually for food, in open net pens, fish tanks, or artificial enclosures such as fish ponds.

Aquaculture actually began thousands of years ago in an effort to provide food security for communities around the world. It uses very little land and water and has one of the lowest carbon footprints of any form of animal protein production. Currently, the aquaculture industry generates less than 0.5 percent of the total global greenhouse gas footprint, while beef production creates 15 percent.[9]

Experts are working to make the climate impact of aquaculture even smaller. As with any other animal production, feeding the fish is one of the most resource-intensive aspects of growing them. Now alternative feed ingredients are being produced to reduce the resource impacts of feed. Other advances in technology and digitization are optimizing feeding practices and efficiencies. Aquaculture also has untapped potential for transparency, which allows us to trace the life of a farmed fish from a carefully selected egg to the point at which it arrives on our plates.

For David, Riverence Trout has provided a bridge from his work as a film and TV producer into the emerging world of sustainable blue foods. In addition to being a storyteller, David is a fisherman who fell in love with salmon in British Columbia. He was mesmerized by this miracle fish that is born in rivers then travels underwater for thousands of miles. Salmon are water migrators: they go to the ocean, collect marine nutrients, and then bring them back to the river. While navigating these adventurous journeys, they feed hundreds of species. Hence, if the salmon goes, so do many other living things.

Of course, salmon are also delicious. David has witnessed firsthand that consumption of salmon is constantly rising, which is why we need a scalable, nourishing alternative to salmon grown the traditional way. David got involved in the search for ventures that protect wild salmon while also providing a sustainable alternative. Aquaculture quickly grabbed his attention as a worthy, needed, and promising concept that can preserve biodiversity in waters, maintain local economies, and provide healthy protein to humans.

In 2017, David raised the bar by getting directly involved in supporting aquaculture in an effort to decrease the pressure on wild salmon. He went out and bought a trout farm. He named this project Riverence, which carries a powerful symbolism referring to one of the most important conditions for protecting rivers and nature: reverence.

Earth's rivers don't run wild anymore. Eighty percent of them are modified for human purposes. This creates a huge responsibility that we must shoulder. If we don't step in and endeavor to protect and save salmon somehow, we might not have them forever. Aquaculture is the best way to take pressure off fishing in the wild and thereby protect and save something bigger than ourselves.[10]

Fast-forward: today, David owns and manages twenty farms, most of them located in the pure, cold rivers of Idaho, harvesting some eight to nine million pounds of trout a year. Supported by a team of scientists, sustainability advisers, fish feed and welfare experts, chefs, and nutritionists, David is on a quest to support nature conservation while raising healthy food and employing local communities. Riverence is working to constantly perfect the trout's feed and nutrition, protect their welfare, minimize waste, and integrate practices to become more and more sustainable.

Each and every Riverence fish has a fully traceable story that the team manages from the very beginning, starting with their own brood stock (eggs) and continuing through egg hatch, grow out, and processing. Without growth hormones, preventive antibiotics, or genetic engineering, the fish is raised through Riverence's own elite pedigree brood lines with decades of proven performance. They are fed a diet rich in astaxanthin, which is the seaweed-based micronutrient that gives ocean-run salmon their rich red flesh color, without requiring the ocean.

David's land-based, raceway (or flow-through) system is borrowing water driven by the powerful nature of Idaho's canyons. But David believes that the blue foods revolution can happen anywhere: in salt water, fresh water, open-ocean pens, or closed containers; inland, in places far away from big natural water basins, in artificial enclosures, and even on the premises of former agriculture farms.

MOVING FROM TRADITIONAL AGRICULTURE TO RESPONSIBLE AQUACULTURE

We are beyond excited to observe the blue foods revolution already happening—including seeing that generational agriculture farms are closing down and being transformed into aquaculture farms.

That's the story of Simply Shrimp, founded by Paul Damhof in the town of Blomkest, in the middle of Minnesota's farm country. In the beginning, no one who knew Damhof could understand why he had ordered twenty aboveground swimming pools. Even his bank found it questionable—they flagged the purchase and called him to verify it. Amazon wouldn't guarantee the pools because of their intended usage, and FedEx double-checked to make sure the delivery was for real.

But eventually the pools were installed on Damhof's family farm, where his grandfather had raised beef cattle and where his father later operated a dairy farm. Today, it's among a handful of operations in Minnesota developing the practice of inland saltwater aquaculture.

Shrimp is by far the most popular seafood eaten in the United States. Americans eat nearly twice as much shrimp as salmon, tuna, and whitefish. Once considered a luxury item coming from the Gulf of Mexico, most shrimp is now imported from Asia at a cheaper price—one that has not only given us "all-you-can-eat shrimp for $5.99" but has also caused catastrophic environmental degradation throughout the world.

Conventional Asian shrimp are farmed in tidal pools. Unfortunately, coastal mangrove forests are often cleared to make way for them, removing a natural storm-surge barrier against typhoons, a highly effective sink for carbon, and a habitat for other species of fish. If poorly maintained, as many are, these pools have a limited lifespan before conditions for farming become untenable. In addition, the shrimp industry has had documented social impacts, including forced labor and outright human trafficking.

Furthermore, traditional wild shrimping is being affected by ocean warming and acidification. These conditions are detrimental to both the shrimp's spawning patterns and their ability to grow transparent shells to avoid predators. With uncertain oceanic conditions and the ethical and environmental challenges surrounding imported shrimp, an opportunity presented itself to Minnesota farmers. Why not take existing big agricultural operations and turn them into blue food beacons closer to the market?

Damhof retrofitted his former calf barn, where it is a consistent 86 degrees Fahrenheit and humid no matter what the outside weather, using a steel skeleton of PVC and hot-foam insulation. The barn is now a place where shrimp smaller than an eyelash arrive. Within 120 days, after being fed a diet that includes squid, bloodworms, and krill, they grow to twenty-shrimp-per-pound size, ready to be scooped from the kiddie pools and sold to order for $20 a pound.

The local government regulates shrimp farming strictly. "Every new batch of shrimp we get has to be preapproved by the Department of Natural Resources," Damhof says. One of his goals is to have zero water discharge. "So we simply reuse the water. The older the water, the better the feed efficiency, and a faster rate of growth makes it easier to manage. Our water is now ten months old and has undergone a 180-degree change from when we started."

These operations have provided a once nearly bankrupt dairy farmer with a way to sustainably contribute a valued blue food into a local food system with low inputs, maintenance, and costs. At a point in history where ocean-based wild stocks of shrimp are endangered by the effects of a warming climate, Paul Damhof is meeting the demand in Minnesota, come snow, rain, or shine.[11]

HOPE IN THE WATER

As earnest blue food ambassadors, we truly believe that the responsible usage of our waters can provide humankind with a long-term source of life, sustenance, and nutrition. Feeding people while protecting the planet is the mission of Fed by Blue, a multilayered project that we started in March 2022. Its goal is to transform blue food systems, empowering people to make responsible blue food choices through transparency, education, and improved policies and practices.

Recognizing and highlighting responsible blue food producers is paramount because they truly prove the impacts of their endeavors, not only for the individuals involved but for the whole community. Inspired by science, personal stories, and nurturing recipes, we invite you to join us in choosing responsibly sourced blue foods that work for both the planet and its growing population.

What we put on our plates today has an impact on our tomorrow. Together, we can set a place for responsibly sourced blue foods at our table. And there are plenty of other opportunities to get involved. From keeping an eye out for responsibly sourced fish and seafood in stores and restaurants, to trying your first kelp chip, to getting involved in policymaking to support responsible production of blue food in our coastal waters, there are many ways we can all make a meaningful splash as good global citizens.

12.

From Lab to Table: The Woman Who Knows Too Much

By Larissa Zimberoff

Larissa Zimberoff is a freelance journalist and author covering the intersection of food and technology. Her new book, *Technically Food: Inside Silicon Valley's Mission to Change What We Eat*, explores how what we eat is rapidly changing and the start-ups behind it. Her work has appeared in the *New York Times*, *Wall Street Journal*, *Bloomberg Businessweek*, the *Atlantic*, *Wired*, *Fast Company*, and more. After working in high tech for a decade, she earned her MFA from the New School in New York City. She lives in the San Francisco Bay area and will gamely try any new food.

My career as a food reporter has been centered on what I call "New Food." This is food created by scientists, not farmers—more "lab to table" than "locally grown." I was a passionate food lover when I began covering the ways technology was reimagining our meals, but after a decade of digging—aka reporting—I felt burdened by the reality of what gets on our plate.

What is all this stuff? I've dedicated myself to trying to answer that question. In New York, I tasted a meal-replacement bar made of spirulina and chlorella—types of algae that you might find at the bottom of a pond. (It turned my teeth green.) I tried insect cookies made with cricket flour. (They were dry.) I found a lettuce farm in what was a former dance club. Black lights and megawatt LEDs propelled the plants to grow. (Bland.)

In the San Francisco Bay area, New Food was everywhere. In Berkeley, I chewed on lab-grown chicken. (Great texture, but flavor and fat? Absent.) In Albany, I bit into bacon brewed with red seaweed protein. (Delicious! Because, well, you know, bacon.) In San Francisco, I pinched cultivated salmon with chopsticks. (A barely there fishy scent and Jell-O-like texture.)

On outings with friends, I'd attempt to describe it all. What I saw. What I tasted. Arms waving, hands gesturing, voice octaving up. Unfamiliar words tumbling out of my mouth. I'd try to simplify. A passerby might think I was enthusiastic. In a way, I was.

A confession: I love getting these first tastes of New Food. I wanted every PR person employed by a New Food start-up to email me with advance news about product launches. Each missive promised me a "first-ever" something or other. I met all such claims with skepticism. I'd ask for clarification and specifics. Sometimes they'd come through, and I'd receive a fancy package in the mail stuffed with products, stickers, and a canvas bag. Other times, they'd ghost me. When I did try the new whatever, I would judge it ruthlessly for the taste, ingredients, and clarity.

I'm not the only one "beta testing foods," as food historian Nadia Berenstein called it in an essay in *Grow* magazine. We're all doing it. Impossible Foods and Beyond Meat are on the menu virtually everywhere we dine. If you fly first-class on United Airlines (anyone? anyone?) you can order the Impossible meatball bowl on domestic flights over eight hundred miles in the continental United States. Plant-based, yes, but highly processed with a long list of ingredients, including additives that may one day prove harmful to our bodies. (Maybe we already have an inkling they are?) Millions of people like me are wondering: What are we to make of this New Food? Is it really good for us? Why or why not? And how are we supposed to decide?

The fact that so many of us are guinea pigs for the experiments being conducted by New Food founders has given me a front-row seat and a privileged place in the foodie hierarchy. In her essay in *Grow*, Berenstein called me "one of the world's few connoisseurs of future food in its beta-testing stage, having sampled everything from mycelium-based steak and lab-made cocoa-free chocolate to cream cheese coaxed from microbes found in geysers at Yellowstone National Park." (You can watch me on TikTok extolling the wonders of that microbe cream cheese. The spread was pretty good despite its subtle "Uh, what is this?" I'd buy it, I said in the video, because these days I'm eating less dairy. When I do eat dairy, I want it to be the stinkiest and gooiest of soft cheeses.)

Over time, my need to learn more about how and what I should eat merged with a widespread desire on the part of countless readers to know more about what they are eating. Eventually, these aspirations gave rise to a book. *Technically Food: Inside Silicon Valley's Mission to Change What We Eat* came out in 2021. I think of it as my "pandemic baby."

So, having devoted a serious chunk of my life to exploring the New Food phenomenon, where do I come down on it? Am I a fan of New Food? A supporter? A critical journalist? Could I be them all?

After several years of sampling and learning, thinking and chewing, I grapple with a convoluted future comprising the whole foods, processed foods, and junk foods we already know plus a road ahead that's paved with a new subset of stuff we don't fully understand. Stuff that has only been grown in a lab for a few years. Stuff that wasn't eaten five years ago. The scientific jargon is a hassle to understand. And transparency—key to organic and sustainable agriculture—is usually absent. The options are daunting for anyone who doesn't want to read my book, look closely at labels on food, and investigate complex words.

Even one of the most basic human foods, milk, needs a tour guide with a PhD. Right now, you can buy versions of milk made from oat, hemp, sesame, flax, soy, rice, almond, green banana flour, mushroom powder, and yeah, even a cow. Very soon, your grocery store will stock milk made from the same protein found in cow's milk yet not originating from an udder. It smells and tastes like cow's milk, which may come as a surprise. It surprised me, even though I ditched dairy milk long ago. This new milk that's not milk-milk will be industrially made but not come from industrially raised cows. It will include a variety of ingredients like emulsifiers, gums, and fats to give it the same mouthfeel of milk and keep it from separating in your morning coffee. Sounds . . . delicious? If it tastes, smells, and feels like milk but comes from a lab, what should we call it?

In this ever more complicated world, how we make our daily eating decisions is increasingly stressful. If we care about the climate, we might pick the milk with the lightest greenhouse gas emissions, right? If we want what we know, we can choose traditional milk. If we're tracking our health, we may opt for plant-based, because that's what the headlines seem to be touting. Many of us shop based solely on the lowest price. What foods do we purchase then? The lowest quality from industrial origins? How do we juggle our varying and evolving priorities?

My focus on how food is manufactured, the science behind it, and what the ingredients are comes from a lifetime of watching what I put in my mouth. I have type 1 diabetes, which I was diagnosed with when I was twelve. Diabetes, an autoimmune disease, is a heavy lift. It requires near constant attention.

Mindful eating is an aspirational goal for many. When I watch what I eat, I'm rewarded with "normal" blood sugar numbers—roughly 80 to 120. This means I can pay less attention to my condition. (I don't like to call it a disease, but it certainly is one.) Health is my North Star, its urgency amplified by the fact that my body doesn't produce insulin, a crucial hormone that breaks food down into glucose that fuels every basic human function. Without it, I can't eat. Quite the dilemma for the girl who loves food.

To address this, I spent my thirties and forties tinkering with my food choices. I used my body and the continuous stream of data from my glucose monitor as a feedback loop to test new diets, foods, exercises, and sleep. The headline discovery: simpler was better. The fewer ingredients with the lightest processing was by far the best. Yes, I'm talking fruits, vegetables, grains, legumes, seeds, and nuts. These are "whole foods," a phrase pulled from the term "whole foods plant-based" that was coined as a way to refer to vegetarian eating. It's the healthy core around which I build my own way of eating. My diet also includes some dairy, some fish, and, far more occasionally, eggs and meat. At my desk in the morning, I drink decaf coffee and green tea. So that you don't think I'm some utopian eater, I enjoy wine and dark chocolate, and when I go on a long hike, there's a scone in my fanny pack. For me, thinking about what I eat is an everyday obsession.

When I visit the grocery store, my personal priority is health. Taste is a close second. My ongoing concerns, and a writing assignment from Consumer Reports, led me to interview Dr. Josephine Connolly-Schoonen, director of the Nutrition Division in the Department of Family, Population, and Preventive Medicine at Stony Brook Medicine in New York. Her title was a mouthful, reflecting a lifetime of study in a range of interrelated scientific specialties, and I eagerly threw every question I had at her.

"What's your own diet?" I asked.

"I'm plant-based but not vegetarian and very whole foods," she said (referring to the movement, not the store). She emphasized that whole foods are more important than being strictly plant-based. "I'll add meat sometimes. I'm short, I'm old, and I don't need a lot of calories." A funny nutrition doctor. She was right.

I mentioned that I used almond-coconut-milk creamer in my morning tea. "Is that okay?" I asked. She said I could find "acceptable plant-based creamers" with "minimal ingredients" but that most have emulsifiers for mouthfeel. These emulsifiers aren't new to processed foods, but our reliance on them is steadily increasing with New Food. "There are some indicators that this could be bad for our GI [gastrointestinal] tract—specifically the integrity of the lining." The research is limited for now but worth paying attention to. I wrote the warning down and mentally underlined it.

New Food will require knowledge of a complex web of information. For me, it's fascinating research. I have the luxury of taking my flood of questions to experts like Connolly-Schoonen and often to the founders of the start-ups that are launching New Foods at such an amazing clip. But we can't all ring up the experts and insiders when we have questions. Pros and cons will need to be weighed. It's exhausting, and becoming more so.

"Somehow this most elemental of activities—figuring out what to eat—has come to require a remarkable amount of expert help." Michael Pollan wrote this in 2006 in *The Omnivore's Dilemma*, which is still pertinent to truly understanding how the food we eat is being made today. Silly as it may sound, what he thought at the time was a remarkable amount of expert help—say, one expert—today might require more like five experts. And it's only going to get that much worse tomorrow.

The direction our food system is taking concerns me. We're seeing more processing, not less, and often a greater degree of obfuscation of food origins. As a result, my long list of unanswered questions keeps getting longer. Can these novel ingredients and methods sustain humans in beneficial ways? Will we truly thrive on analogues—things that are "just like" milk, or "just like" whatever—when the original foods are no longer widely available or affordable?

I've now interviewed hundreds of New Food founders. What's in their crosshairs is usually the same. There's a list of serious problems impacting our food system that they intend to address: climate change, animal welfare, the depletion of resources like soil and water. Solutions, they assure me, will fall into our laps if they can only raise enough money to scale their new versions of milk or meat. When and how protein comes into existence is where their sense of urgency comes from.

But what about the mind of the consumer? During the pandemic, people began considering their bodies more. We're starting to realize, it seems, that our health begins with what we eat. Yet, globally, we seem to be suffering from more diseases, not less. It's widely understood that eating less meat would produce enormous medical benefits for most people in the developed world. Yet few of us have gone completely vegetarian; even fewer have gone vegan. How do we change that?

Cell-based meat, also called cultivated meat or lab-grown meat, is one of the New Foods working its way into the mainstream. In 2016, after raising almost $4 million from investors, Upside Foods, a cell-based meat company, turned to crowdfunding site Indiegogo. The campaign allowed Uma Valeti, cofounder and CEO, to gauge the reaction of techy insiders, first adopters, and other potential consumers to his question: "Should we grow meat from animal cells rather than raising an animal?"

In January 2020, I interviewed Valeti in his Berkeley offices. "There was an enormous amount of interest," he said. "People were actually willing to give something a try, even when it wasn't available." (Kinda like the early days of Elon Musk's Tesla.) Almost three thousand people pledged money to try this imaginary meat. The money raised by the campaign was a paltry $136,677, but the reaction reassured Valeti that "believers" were out there—that these people would eventually be his consumers.

Today, Upside Foods has raised north of $600 million and is valued at over $1 billion—a number that I expect to soar. In 2022, Upside's cell-based chicken received a "no questions" letter from the FDA, which means it's approved as safe for human consumption. The start-up took out a full-page ad in the *New York Times* to celebrate. (The ad spoke directly to chickens, letting them know they were "dismissed from the dinner table.") Government validation is a huge win for an industry that was once seen as pure science fiction. It will take time, years if not decades, before you see lab-grown chicken at your local Safeway or on the menu at McDonald's, but it's one step closer to your plate.

Cell-based meat isn't yet a sure thing. The attempt to scale it to feed more than a few fine-dining patrons will take a world of scientific effort and inordinate amounts of capital. The hurdles are many and include how to reliably feed cells, what the nutrients will be, how to grow trillions of cells without fail, how to meet energy and water needs, and how waste will be disposed of. Farther downstream lurks an even bigger question: What will consumer acceptance be? As the food-tech industry focuses on solving those problems, consumers will wait and wonder. What are the unintended consequences of shifting our diets and food economies to animal protein that began its life in a petri dish and not from an animal that grazes the land or swims the sea?

I won't ascend a soapbox to say that we need to stick to legacy foods we have always eaten (spelt, anyone?) or to assert that there is one perfect way to nourish our bodies. However, I do expect these visionaries to be considering health holistically, as the output of a system made up of a multitude of components in a single day. There is no silver bullet to fixing what ails our food system, and this includes a single solution like lab-grown meat. Will it be healthier for me? Will it be more delicious? Will it be less expensive? Will it help save our beleaguered environment? Or will the energy and nutrient needs of lab-grown meat make it less of a climate win? I have an endless supply of questions for founders who are oh-so-certain that their product is the solution to a big problem like climate change or land or water use.

The last chapter of my book is probably the most compelling. In it, I asked experts what we will be eating in twenty years. Lisa Feria, managing director of Stray Dog Capital, assured me that cell-based meat would become mainstream in twenty years. Chef Dan Barber envisioned seeds of the future bred for better flavor and hyperlocal regionality. Serial entrepreneur Mark Cuban desired a "low-cost cube that's satiating, tastes great, and fulfills the recommended daily allowance (RDA) of nutrients for less than one dollar"—his answer to the problem of food insecurity. Writer and designer Ali Bouzari foresaw investors putting millions into making a really great sweet potato. Journalist Tamar Haspel extolled grains and legumes taking over the center of our plate—not meat. And finally, author Vaughn Tan brought it all back to reality by reminding us that nature is more creative than we can imagine: "In order to simulate naturally great food that surprises and delights, the key thing we need to learn to replicate is [its] unpredictability."

Unpredictability is not what scientists hope to create.

This chapter offered me so much hope. There will be moments of beauty in our uncertain path ahead. Yet, as I look out my office window, these visionary beliefs are muddled by what I've learned. All my visits to labs. Colorless, sometimes odorless prototype foods I am handed to try. Nondisclosure agreements I am made to sign before I can see how the food is being made. I feel like the Woman Who Knows Too Much. Being on the front lines of system transformation is thrilling. But the feeling is overshadowed by my ambivalence regarding the promises of future food. This is hard for me to admit.

New Food seems to be trying to do too much. Founders have told me their plant-based burgers will replace so many cows that it will solve climate change. Other founders tell me their labs will make a facsimile product, like chocolate, that can scale without impacting the environment or taking jobs from farmers. Still others tell me they will grow fish in a lab, thereby replacing our abused and overfished oceans. These are all ambitious solutions aimed at solving big problems.

But can these processed, made-in-a-factory New Foods sustain the body as completely, and as safely, as the farm-grown kale or the chicken-hatched egg they replace? How can they? How can I, the consumer, know that this New Food has the same macro- and micronutrients as the nature-made original? And how can anyone know that a New Food that has existed for only a few years will be safe to eat over the long term?

Of course, we all want to make a dent in the big problems that the New Food founders want to tackle; the passion and creativity they bring to these issues is inspiring and important. But how can we—as a community, as consumers, even as investors—possibly know that we aren't creating as many new problems as we are potentially solving?

Looking back on my years as a New Food journalist, what's clearer than ever is the kind of future I want for our food system. I want people to think about what they eat. I want people to shop for the climate. I want people (including me!) to exercise more. I want people to eat more plants and less meat. I want an end to industrial animal agriculture and inhumane, pollution-spewing feedlots. I want companies to stop selling cheap, ultra-processed junk food. I want an end to food deserts. I want public health policy that ensures every single person can eat nutritious foods. I want the government to limit how much crappy food Big Food is allowed to sell and put an end to escalating profits because we can't stop buying what they're selling. I want the FDA to have more resources to watch over the evolution that may come from food tech. Phew. It's a lot.

I also want investors to believe in ideas that don't hinge on proprietary technology. I want a future that addresses the past and engages with it in a creative, renewing way rather than discarding it. I want it all—and I'm hopeful these wants can eventually be converted into healthy food on my plate about which I will proclaim loudly, "This is delicious."

13.

Healthy Food for All Our Kids

By Nancy Easton

Nancy Easton is the executive director and founder of the national nonprofit Wellness in the Schools (WITS). Through meaningful public-private partnerships, WITS empowers schools to provide healthy, scratch-cooked meals, active recess periods, and fitness and nutrition education. She has been honored by First Lady Michelle Obama at the launch of Chefs Move! to Schools, named a Food Revolution Hero by acclaimed chef and food activist Jamie Oliver, and recognized by Ann Cooper, the "Renegade Lunch Lady," for her dedication to school lunch reform. Before founding WITS, Easton was a teacher and school leader with the New York City Department of Education. She holds a school administration certificate from Fordham University, a master's from Bank Street College of Education, and a bachelor's from Princeton University, where she was a three-sport athlete.

For two decades, I've been working with allies across the United States to make sure that every child has access to healthy food and active play in their school.

I approach this work through a social justice lens. I was raised in Miami, Florida, a very diverse yet segregated city, and at a young age I struggled to understand the many disparities I saw before me. Every day after school, my brother drove me and my track teammates home from practice in our father's brown, hand-me-down Peugeot, with a speedometer that didn't exceed eighty. We would drop off my Black friends and teammates at their homes—the housing projects not too far from school—and then continue on our way, over a bridge, to our all-white beach community. I didn't quite understand why our lives were essentially separated by the color of our skin.

In the early 1980s, when our public school was canceled due to nearby race riots, those of us living on Key Biscayne spent the days off from school waterskiing in Biscayne Bay, while across the ocean, homes, businesses, and lives were being destroyed—including the homes of my friends and teammates. We could literally see the smoke from fires burning them to the ground. A body of water separated Black from white, the haves from the have-nots. We were a solidly middle-class community whose skin color gave us much more privilege. And I was confused.

My personal experiences mirror the systemic realities that have shaped and permeated our country. Social justice and food justice are inextricably linked in a set of patterns we now describe as social determinants of health, involving the conditions in the environments where we live, work, and play that impact our health outcomes. Research shows that, due to systemic racism, Black people are more likely than whites to live in poor neighborhoods with fewer social services and less access to healthy food.[1]

So now I work to address disparities in food access. I do this work because of what I witnessed as a thirteen-year-old who could not make any sense of it. The injustice still seems nonsensical, yet at least I now understand why it exists and how I can perhaps do something to make a difference.

As a young school leader in New York City in the early 1990s, I saw the impact that poor diet and a lack of physical activity can have on children's ability to learn. I started a school with two of the first employees of Teach for America, the inspiring nonprofit organization founded by Wendy Kopp. The three of us came together wanting better circumstances for children in NYC public schools, children whose families lived below the poverty level. We believed that a good education was the way to a better future for them. Just teach children how to read, how to write, and how to do math, and they will work their way out of poverty. We were so idealistic, and it seemed so simple at the time.

What we didn't yet realize was that, while we were planning to rock the system, we were still working *within* a public school system that was (and still is) so broken. We were fighting systemic injustices that went well beyond our school walls and that were a true microcosm for the world at large. Gradually, we learned that reading, writing, and arithmetic were not the panacea. We learned that children first needed to be nourished, in all the many meanings of that word.

For me, it started with health.

I had the wonderful responsibility of monitoring school breakfast and lunch, a favorite job of mine, because it let me connect with middle school students in ways that I couldn't in an academic setting. Apart from getting to know the kids, what I witnessed during breakfast and lunch was dismal. For breakfast, my middle school students usually arrived with a black bag from the corner deli filled with Cheetos or Doritos and a bottle of orange soda. Not exactly the breakfast of champions. Lunch was an overly processed school meal, and recess was sedentary. I often pointed out that not only did the students have orange fingers and orange tongues for a good part of their day (from the chips and soda), but they also could not walk a flight of stairs without pausing to catch their breath; they couldn't focus in class. Without their health, they couldn't learn.

Back in the early 1990s, not too many people were talking about the obesity-related illnesses that begin in childhood: heart disease, diabetes, cancers, strokes, dementia. But I was watching the problem take shape before my eyes, and I decided to do something about it. To help you fully understand what "it" is, here are just a few data points:

- More than 40 percent of adults in the United States are obese today, up from 5 percent in 1960.
- One in four teenagers in the United States has prediabetes or type 2 diabetes.

- One in five NYC kindergartners enters school already obese and at risk of disease.
- The United States spends nearly $200 billion each year on obesity-related illnesses such as heart disease, diabetes, cancers, strokes, and dementia.

At first, I wasn't exactly sure what I could do, but I started with what I knew. I knew schools, and I deeply believed that schools have the power to transform—not only a child, but a system. I also knew that, when it comes to the health of schoolkids, NYC feeds and plays with a million children every day. What an awesome responsibility and opportunity to make those meals and those playtimes the best they can be. That's where I decided to start.

Having started working in a new school, I decided that I would demonstrate exactly how to serve the perfect school lunch. My daughter was in a class of eighteen three-year-olds, an incredibly diverse group of children whose parents included teachers and administrators as well as high schoolers who attended school in the same building. The children would typically go to the cafeteria with their teachers at lunchtime, then balance their trays on the walk back to their classroom to eat in a quiet setting.

I thought, what a perfect place to begin changing school meals in NYC! We would take only fruit and milk from the cafeteria and leave behind the chicken fingers and french fries. Meanwhile, I had engaged volunteer chefs to prepare scratch-cooked meals for the children in a makeshift kitchen three days a week. I also convinced a local vegan restaurant and a healthy pizza joint to donate meals on the other two days. It worked for a while. But periodically, the volunteer chef would get a paying job (good for her, bad for me), and I would find myself—highly pregnant at the time—preparing school lunches while squeezing awkwardly into the space between the open oven and the wall behind me.

I managed to make it through the school year, but I realized that my little experiment was not exactly going to produce any systemic shift in school meals. But I was determined to keep going.

The following year, when the same daughter moved to our neighborhood public school, I began poking my head into the cafeteria and garnered the attention of the school food Manhattan director for the NYC Department of Education. I was so excited when he wanted to meet with me—until he met with me! Today, that gentleman is one of my most important advocates and partners. But at the time, he came to give me a little talking-to. He wondered why I was interested in rocking the status quo. He thought that maybe it was best for me to keep my concerns and big ideas to myself. He had no idea!

Instead, we overhauled the salad bar. We cleared out the pickles, onions, and mealy out-of-season tomatoes and replaced them with a bounty of fresh lettuces, a rotating variety of at least three other raw vegetables, a selection of homemade salad dressings, and a mixed salad of the day. We engaged parents to encourage children to help themselves from the salad bar and help the younger ones as needed. Soon we began to talk about other changes on the menu itself. I quickly realized that this was getting to be a bigger project than a few overworked but dedicated moms could handle.

Enter Chef Bill Telepan, a parent at our local neighborhood school, as well as the chef and owner of a popular Upper West Side and Michelin-star-studded restaurant, also named Telepan. One evening, we moms and our new best friends from the school kitchen—the cooks and the manager—were showcasing our newly developed scratch-cooked and delicious school lunch menu items for open school night. Bill came for his daughter's parent-teacher conference, at which, as he repeatedly and proudly shares, his second grader scored high marks. More important, Bill grabbed a few chicken Caesar wraps from our table on his way in, and on his way out, he asked me, "How can I help?"

Those four words truly shaped the organization that became Wellness in the Schools.

Bill and I got to work. We knew a few moms who were in culinary school and together decided to engage these moms as support chefs in the school kitchens. We expanded from our children's school to two others, located in communities with less access to the resources we enjoyed. As I have described, poverty and health are linked, and I always wanted to drive larger systemic changes beyond my own kids' schools. It is worth noting that I was also well aware of the great privilege that working in a city like New York offers to the type of model we were developing. With one email, Bill could gather a handful of peers to adopt schools and replicate our programming with relative ease. However, our growth is attributed not to celebrity chefs but to hardworking, passionate change agents who reside in communities all over this country.

We began to grow, one school at a time, starting with two staff members and forty volunteer chefs. By 2008, our Cook for Kids program was in eight NYC schools. Each volunteer chef would take a day off from their paying job to work in a public school kitchen: five volunteer chefs, one each day of the week, eight schools. We introduced scratch-cooked recipes that the chefs helped to prepare and began a fledgling side-by-side training program in school kitchens. Each evening, the chef would email their team to remind them what had been prepared that day and what remained for service for the next day's team. A few examples:

Well, another busy day at PS XXX. This morning, after discussing who was going to win *Dancing with the Stars* (I watch with my kids), we prepped a full salad bar. We have been working on fridge organization (FIFO). Miguel has been keeping the walk-in beautifully and it's so easy to find all the veggies. He also works in a SoHo restaurant. It was suggested to me that we be careful when working with the tomatoes to make sure there is no brown on them. The kids are picky and like their tomatoes red.

Kaci, there is another container of cucumbers ready for tomorrow.

Good Morning. Yesterday I made a spinach salad with carrots. Mary and I cut up/sauteed several tomatoes and mushrooms and mixed with tofu. In addition, I made a broccoli and onion side dish.

There is a plentiful amount of romaine lettuce to be used as well as tomatoes and some peppers.

I think the youngest group of kids here are definitely eating more from the salad bar. Granted you have to convince them, but they seem to leave with a smile on their face. It kills me to see their plates filled with processed/fried foods.

[Head school cook] Kenny is applying for a cooking position at various schools, so he was not in the kitchen yesterday. I wish him luck but also think he brings some spirit to this kitchen :)

I waited anxiously to read the notes each night and to revel in the difference we were making in those eight schools.

Looking back, I am struck not only by the big impact that a small grassroots organization can make but also by how the power of "people talking to people" is critical to making change. Early on, a colleague sent me a *New Yorker* piece that had absolutely nothing and absolutely everything to do with the work that we do. It was a piece on childbirth in developing countries and a research project that highlighted the importance of hands-on support in making real change. It's the same with school food. I always share with my team that they will be much more likely to teach a school cook how to best shock broccoli if they also know what that school cook is carrying with him or her that day, whether it is excitement about the upcoming baptism of a grandchild or the stress of rushing to their next job to make ends meet. And this has become the backbone of WITS: people talking to people.

We learned a few other things from that eight-school year. First, that there is an extraordinary volunteer workforce in NYC, but it is not sustainable. Second, that we needed to introduce an educational component if the kids were actually going to eat the food. And third, that we needed to grow as an organization if we were going to make any kind of significant impact.

I am certain that there were one trillion more lessons and actions needed to be part of this systemic change thing, but those three takeaways were the most poignant at the time—and perhaps all we could handle.

I pounded my fist on the table and said that if we were going to make an impact, we had to expand substantially. Not to just two more schools, or to five, but to…twenty. Somehow twenty seemed big and like a good number. But keep in mind that there are 1,200 school buildings in the great city of New York.

Thus began the process of raising money to pay a workforce. A donor with a big name gave us $25,000 and suggested that we leverage their funds (and their name) to get others to do the same. I had to do that nineteen more times to raise what we needed. While this was daunting for someone who had zero fundraising experience, I so deeply believed in what we were doing that I was determined to get the job done. At some point, as September and the upcoming school year was approaching, we had enough funds for nineteen schools. I was getting anxious and declared that nineteen was enough.

From nineteen schools, we grew and grew. In 2023, we are creating scratch-cooked recipes, training school cooks in their preparation, and educating kids, adults, educators, and communities about *why* healthy eating is important in five states, ten school districts, and more than 120 schools, serving sixty-two thousand children. (That number is actually down from the years leading up to the COVID-19 pandemic, when we were serving nearly one hundred thousand children. We are now rapidly bringing our work back to that scale and aiming for much more.)

The Green for Kids program is an extension of Cook for Kids. Our goal is that all schools we work in have a school garden, which may range from a small window box to an edible schoolyard.[2] We partner with several organizations to bring gardens to schools; then the role of the WITS chef is to bring it to life by harvesting and working with the kitchen staff to serve on the line, typically through a "garden to café day."

From the very beginning, we also knew that eliminating obesity-related illnesses that begin in childhood would require a focus not only on food but also on fitness. The Coach for Kids program evolved almost in parallel with Cook for Kids. We began placing coaches in the recess yard early on. They were charged with "getting the least active kids active." As a result, school lunch and recess work hand in hand: children eat lunch, go to recess, and then go back to class ready to focus and learn.

Our WITS coaches serve a very similar role in the recess yard as our WITS chefs do in the school kitchens: they create change. They make sure that kids have established areas in which to play, they train school aides (the recess yard staff) in play methodologies, and they provide a tool kit of new games and activities for them to implement. Most of all, they play with kids. They often act as the quarterback of the football game or the head cheerleader at the box-ball court. And, yes, our coaches also become role models for the students and advisers for the school aides.

To impact a greater number of kids on a daily basis, we are growing our coach program to include a live virtual component, whereby students can engage directly with a coach who "joins" their class in a daily fifteen-minute fitness break. This program also provides a library of breaks for teachers to tap into at any point in their day when they need to refocus their students.

Our models have evolved from the one flagship program to include intensive, up-front training programs (CookCamp and BootCamp) as well as workshops and consultancy programs. This allows us to meet the growing demand for our work without exhausting our resources. In these programs, we train the adults or provide workshops to enhance skills, all the while driving home the *why* of the work. We partner with school districts to provide nutrition and fitness education, healthy scratch-cooked meals, and active recess periods. Trained culinary graduates partner with cafeteria staff "to feed kids real food," and fitness coaches encourage schools "to let kids play." Our approach improves student outcomes and drives systemic, long-term change, shifting school cultures and ultimately fighting the childhood obesity epidemic.

Today, our weekly kitchen notes look more like these examples:

> While in the kitchen at PS XXX, I was able to see a number of ways that the head cook incorporated what she'd learned at CookCamp in October, which was another satisfying and joyful, frankly, part of my week. There were creatively composed salads on the Monday return from Thanksgiving (quite a feat in itself), scratch croutons on the chicken grab-and-go salads, and charts of

responsibilities for the cooks. She was also wearing her WITS bandana, so there's extra points there.

So many powerful moments occurred over the last three days. There were quiet moments of connection and shared moments of understanding; outward expressions of support and validation; little smiles and big laughs; cheers and applause. And a lot of "thank you for that." I felt so grateful for every one of them, whether as a witness or participant, or hearing a story shared by one of you.

You were all an integral part of the last three days and it could not have taken place without you. We don't get to spend very much time as a collective unit, and each time we do I think we all walk away with hearts full. This week we were able to share that magic with OFNS [Office of Food and Nutrition Services] cooks.

Thank you, each of you, for showing up and jumping in. You all added something of great value as we continue building the framework to make Healthy Bodies = Healthy Minds.

We were people talking to people. The very best way to create change. What has become most poignant for me over the decades is the relationships our WITS chefs and WITS coaches build with each other and with the school communities. When I first started Wellness in the Schools, I modeled it after Teach for America, where recent college graduates are hired to work two years in a school, eventually moving on to a long-term career. Our WITS chefs and coaches were hired under this same premise. But as it turns out, we have team members who have been with us for over ten years.

Perhaps the reason is that the work is so far-reaching and compelling. It's more than teaching cooking, nutrition, and fitness; it's more than teaching culinary skills to school cooks; it's more than helping school aides to run their recess yards; it's even more than helping schools and communities fight an epidemic. What this work has evolved into is what I described earlier: "People talking to people" to bring about change. It's about working closely with communities, about understanding the incredible barriers that most of our schools and families face each day and the lack of access to healthy food and safe spaces to play. It's about bringing a little bit of hope by teaching skills and helping children (and adults) to develop habits that can last a lifetime, for a lifetime of better health.

I have come to understand not only the great power of trained chefs and coaches, the majority of whom are representative of the communities we serve, but more broadly the power of dedicated, thoughtful, and caring humans. We do this work because we care deeply about the future of health in this country. And while every day is not smooth, we are moved every day by the young people with whom we work and for whom our team members serve as role models in their health journey.

We are successful because of our deep attention and commitment to the *why* of our work. When we begin a CookCamp workshop with school cooks, we start the conversation in a circle, asking each participant to share why they are motivated to do their work each day. This typically brings us to a discussion about school meals, the obesity epidemic, and why their roles as cooks in a school kitchen are so important—how they have the power to drive change for so many young children, every day. This is why we work in schools. I do believe that this approach is one of the most important ingredients to our success: the way we show up, not wearing capes but understanding the challenges as much as the opportunities and providing as much support as we do "answers."

Although I often felt discouraged as a young school leader, interestingly I have become less jaded over the years. I have come to understand not only the challenges but the great opportunities that exist inside the walls of a school building. This is the only place where many children have access to meals and a safe place to play. In NYC alone, a million public schoolchildren rely on the federal lunch program. They have access to hot meals and free play every day—all free, for all children. What an opportunity. This is exactly where change can and should happen. That is pretty powerful. And that is why we do this work each day.

Speaking of hope, I will end with what I see in front of us for school meals. In the years that I have been doing this work, it has certainly become a much more popular place to be. In the early years, it was just me and the hippie moms breastfeeding on the playgrounds, talking about organic tofu, grass-fed beef, and locally grown produce. Today, young men and women are graduating from college with degrees in food studies and food policy; start-up companies are developing plant-based delicacies; the fitness industry has gone wild; the White House held its first conference in fifty years on hunger, nutrition, and health; and food companies and hunger organizations are now playmates. It is a different world. And while school meals have a very long way to go, the conversation around the need for change is very real.

The conversation in NYC is so real that our mayor, Eric L. Adams, and his administration invited Wellness in the Schools to bring our training program into all of the city's 1,200 public schools. Beginning in fall 2023, each of those kitchens will have a WITS chef working side by side with the school cooks to prepare scratch-cooked, plant-based, culturally relevant meals for *all* NYC schoolkids. In a new program we call Chefs in the Schools, WITS is developing the recipes, testing them in schools with the real critics (the children), and then training the school cooks on their preparation. We have brought together a Chef Council, composed of local and celebrity chefs and chaired by Rachael Ray, to develop the recipes with us and the NYC Department of Education Office of Food and Nutrition Services.[3] This is all a testament to the momentum surrounding school meals and the desire for change.

It might seem unusual that we are launching Chefs in the Schools smack in the middle of a labor shortage and an ongoing supply-chain disaster, but twenty years later I am the same girl who was told I could not change school meals in one small classroom with eighteen children. Today, I can at least recognize the boldness of this endeavor, without the naivete of my younger self, and I know that, wherever we land, it will be worth it.

These aren't the only new initiatives on our horizon. When the Chef Council garnered national attention, our Coach for Kids partners came knocking on the door seeking a similar program, and we responded. We are now also developing a Coach Council that will provide expertise and professional development for school-based physical education and physical activity instructors.

In a similar hopeful vein, in April 2023, one hundred food service directors from around the country gathered in Austin, Texas, for three days to participate in workshops on how to move their districts to a more scratch-cooked operation. This is an outgrowth of a project launched in the summer of 2019, when I gathered a handful of nonprofit partners working in the school lunch space together with six progressive school districts in NYC in an effort then called the School Food Innovation Lab. After individual interviews with each founding member, we came together in NYC to innovate. We were motivated by a sense of urgency. The small organization that I'd started fourteen years earlier in one school was making incremental impact, but every time I met with a food service director, it was clear that too many barriers remained: funding, staffing, training, education. With support from friends in the space—the Whole Kids Foundation, the Life Time Foundation, and the Chef Ann Foundation—we spent two days rolling up our sleeves, getting messy, disagreeing, agreeing, creating, tackling problems, finding solutions, and then planning how to activate.

From those two days, we emerged as ScratchWorks.[4] It's an initiative by and for food service operators, operating as a collective rather than being led by just one organization, and working within the system to support school districts in their efforts to move to more scratch cooking, regardless of where they start on a scratch-cooking continuum. In the last two years, we have worked alongside a policy partner to introduce a definition of scratch cooking into legislation and to introduce bills at the local, state, and federal levels that would give grants to districts to build their scratch-cooking programs.

That brings us to our April 2023 gathering, at which the founding districts and others offered presentations on culinary methods, ingredient procurement, marketing, and education. More important, the district leaders left with connections to a network of professionals who are working toward the same end goal—to cook more from scratch in public school kitchens—and who will become their mentors. More people talking to people.

My hope barometer for school meals and for school play is quite high. It's driven by the momentum I see around the country. I've offered a few examples in which I play a role, but there is so much more happening. Some days, I feel deflated and exhausted by the effort involved in trying to do something as seemingly simple as preparing a vegetarian entrée or offering up a fitness break in the middle of a school day. But, most days, I am motivated by the many successes that I also get to witness. It took me decades to finally realize and accept that I am actually one of the lucky few who, on good days, can merge my life and work together and feel a great sense of satisfaction.

When I started this work, I had two young children and a newborn. I remember the time my eldest asked me, "Are you going to be the lunch lady in my school? And are you going to make me eat all my kale?"

Today, I believe that he and his siblings recognize that my work is really not about kale. It's about connecting with others to bring about change. I hope they've also learned from me a little something about health and wellness, and a little something more about the value of human connection. I hope they will always dive in with tempered optimism and that, even when they fail, they will keep talking to people—and maybe eating a bit of kale along the way.

14.

The Four Bites: How We Can Transcend Divisive Meat Debates and Nourish Humanity Sustainably

By Christiana Musk

Christiana Musk is cofounder of Flourish Trust, a philanthropic fund dedicated to fostering healing for people and the planet. Her background spans environmental and social entrepreneurship, investment, nonprofit, and foundation leadership. She serves as chairwoman of the board of the Unreasonable Group, a public benefit company that resources impact entrepreneurs. She is the host of the BBC's *Unreasonable Impact: Food Solutions* podcast. For over a decade, Musk has been on the front lines of emerging solutions in sustainable food systems. She led a global multi-stakeholder initiative on sustainable diets and helped develop award-winning research that raised awareness of the climate impacts of livestock. Her food policy research on competing worldviews in meat and sustainability at City, University of London, was awarded for its contribution to British farming. She was cofounder of Zaadz.com and a founding partner of clean energy retailer Green Mountain Energy.

A single bite of duck meat, pan-seared in olive oil, was placed on the conference room table in front of me. "It is actually *real* meat," the tissue engineer emphasized, "but it's also *better* than meat." I leaned forward, punctured the flesh with my fork, brought it toward my mouth, and paused. I inhaled the aroma and took in the popping sound of sizzling gristle that's unique to meat that's freshly seared, crisp, and ready to eat.

I looked around the table, where those around me wore near-identical expressions of excitement tinged with stress. Sitting across from me were the cofounders of the cell-based meat company Upside Foods, Uma Valeti and Nicholas Genovese. More onlookers stood huddled in a tight circle around the table, including several scientists as well as the chef who had cooked the duck breast. As I prepared to take my first-ever bite of cell-based meat (also known as cultivated meat), someone reached out a trembling arm, trying to immortalize on an iPhone my reaction to a bite that's aspiring to revolutionize the global meat and seafood industry—which is expected to grow to over $1.6 trillion by 2030.[1]

Before we sat down at the conference table, I had been given a tour of the lab at a research facility just outside of Oakland, California. I was shown a small sample of cells, previously extracted from a living duck. The cells were then nurtured in growth serum within a petri dish, provided a scaffolding to train their shape and texture, and subsequently shifted into a series of ever-larger containers until the cells had finally bloomed into a breast-like shape the ballerina-slipper-pink color of fresh, raw poultry. The duck breast was harvested from the lab and sent down the hallway to be seared in front of me in the test kitchen and plated alongside some root vegetables and a dipping sauce.

What sat on the end of my fork was nothing less than the promise of tissue that infinitely renews itself to provide abundant meat for all. Meat without pain, suffering, and death. Meat that could satisfy an insatiable global appetite without devouring rainforests and the creatures and native peoples who live there. Meat without climate-changing consequences. Meat that doesn't monopolize fully a third of the ice-free surface of the Earth to produce feed for or graze the seventy billion land-based animals consumed annually by eight billion people and counting.[2]

The sheer scale of global meat production defies the imagination. In terms of biomass, livestock—including all types of cattle, poultry, and swine—are in aggregate the most dominant animals on Earth, weighing in at more than twice the combined biomass of all humans and all other large wild animals—from deer to elephants and everything in between.[3] These livestock produce up to nine times more waste than every member of the human species combined.[4] If that proportion seems shocking, try nearly doubling it. It's estimated that, by 2050, the demand for animal-sourced foods will mean we would need to raise 120 billion animals a year.[5] And with that spiraling demand will come a near-doubling of greenhouse gas emissions, as well as water depletion, pollution, and the decimation of fragile ecosystems across the globe upon which an incalculable number of wild species of plants and animals depend. Studies have argued that human carnivory is the single greatest threat to global biodiversity.[6]

This doubling of demand is driven not only by the nearly two billion new humans who will join our global dinner table but also by the meatification of global diets. In other words, the demand is largely due to the three billion people already present at our global table who are increasing the amount of meat and dairy that they put on their plates.[7] Producing animal-sourced foods for nearly ten billion people using current practices would require more land than exists, leaving no room for nature, and would overshoot critical emission thresholds, making the dangerous impacts of climate change unavoidable.

Ninety-five percent of the global population eats meat. Meat is built into the foundation of our society; into the very language we use every day. The word "capital" comes from the Anglo-French term *chattel*, meaning cattle. The Latin word for "money," *pecunia*, comes from *pecus*, also meaning cattle. The stock market itself descends from the days in which livestock were traded on Wall Street. Meat is prized, even revered, in most cultures across the world; therefore tackling our meat crisis is a prodigious undertaking. Meat is tremendously valuable to us, which makes the quest to try to replicate it through technology so audacious and expensive.

At the time, cell-based meat was nearly $2,000 a bite, making Valeti's decision to bring me into the Upside Foods lab a pricey one. It was the fall of 2017, and at that point only a few people in the world had ever sampled meat grown from the tissue of a living animal and delivered to the plate without slaughter. Valeti wanted to hear what I had to say about this technological innovation.

Why me? Because I've been immersed in the meat crisis for over two decades in a host of different capacities. I've led a task force to tackle the issue and spearheaded environmental nonprofits and food policy campaigns. As chairwoman of Unreasonable Group—an accelerator for impact entrepreneurs—I get to see leading-edge sustainability solutions for food and agriculture.

Twenty years ago, I had never thought about the ecological impacts of producing food, nor how the food I ate impacted my personal health. I didn't think about food very much at all, until at one point I weighed in at 260 pounds. I was so heavy that in my early twenties I had to stop to catch my breath going up a single flight of stairs. But when I became a mother at a young age, I wanted to be active and enjoy life with my daughter. So, I got voraciously interested in learning everything that I could about healthy food and subsequently lost half my body weight in less than ten months.

As I fed my daughter her first bites of food, I started to ask questions such as: Is this food going to nourish her? Does it contain anything harmful? Where did this apple come from? What cow made this milk? What about this meat? The more questions I asked, the more unsatisfied I was with the answers, until it became apparent that the food system was not healthy, not fair, and not sustainable. Since then, I have devoted most of my career to trying to answer the question: How can we nourish all of humanity adequately, equitably, and sustainably? All paths I've walked down eventually lead me right back to the same place: the inevitable conclusion that meat is at once our most wicked and intractable global problem and also one of our greatest points of leverage for creating a better world.

As the climate emergency surges toward the tipping point, we can no longer afford to allow our agricultural emissions to spiral or to expand our food-print by pushing further into tropical forests, drawing on precious fossil freshwater, sloughing nitrogen waste into our waterways, and depleting scarce phosphorus reserves. Urgent action is needed now.

But what should we actually do? Well, let's scroll through YouTube to look to the experts. One popular TED talk given by bestselling food author Mark Bittman convinced me meat was a major driver of climate change, and the only solution was to eat exclusively vegan foods every day until 6 p.m.[8] Another TED talk with millions of views by legendary biologist Allan Savory informed me that livestock was our only hope to reverse catastrophic climate change, and the answer was to increase stocking rates of cattle to restore carbon-thirsty grasslands.[9]

Wait, what? How could two very smart, well-informed experts have come to completely different conclusions? Does that mean we should produce more or consume less meat? Can we consume unlimited animal products, so long as they're produced in a more "natural" way? Or should we put a tax on animal foods to drastically decrease consumption? If we did, would consuming less meat and dairy prove disastrous for global health, leading to nutritional deficiencies, or is a plant-based diet the secret to optimum fitness and longevity?

Will the planet's biodiversity be saved by intensification or regeneration? In other words, is it better for the environment to be land sparing, by ramping up animal production in concentrated animal feeding operations (CAFOs) and allowing land to return to its natural state as wild habitat? Or should we push for land sharing, casting domesticated ruminants such as cows in the bygone role of their ancestral counterparts, like bison, which historically roamed the land in huge herds, pruning grasses, churning and regenerating the soil?

Does salvation lie in technology or ecology? Should we invest in radical new technologies to increase yields of meat, milk, and eggs to feed the burgeoning human population? Or was it the technology of industrialized agricultural and fake food that sparked the environmental, health, and animal welfare crises in the first place? Can we even produce enough meat without technological aids?

That barrage of conflicting questions, of course, raises the larger, overarching problem of why so many experts—who all claim to base their conclusions on sound science and who command considerable followings—disagree so fundamentally with one another.

The key to answering all these questions lies in unlocking the hidden ideologies that inform the science, policy, and even dietary advice touted in the media, popular cookbooks, and restaurants. But what's the point of digging into ideologies? Aren't the facts the facts? If only. Here is the problem with facts.

In 2013, I was hired by a group of philanthropists to lead a task force with this mission: to aggregate the best of what we know about the meat crisis, consolidate the research, inform our world leaders about the environmental impacts of livestock, and identify scalable strategies to reduce meat consumption. I met with scientific experts on the topic all over the world, including the UN Food and Agriculture Organization (UNFAO) that aggregates global data on livestock, issue specialists like Oxford University's Food Climate Research Network, and the authors of many of the major publications on the issue. Based on the environmental data we had aggregated, we then looked to the international think tank Chatham House to conduct qualitative research on public awareness followed by stakeholder engagement on policy opportunities to address the ecological impacts of meat production. During this time, I traveled continuously, presenting what we had learned to influential audiences.

After years of presenting across the world at events like the UN Climate Change Conference and the USDA dietary guidelines hearings, as well as to private philanthropy groups, leaders of the big environmental nonprofits, and the media on the front lines of the science and policy influencing the meat debate, I became exhausted by getting up onstage and fighting others with conflicting facts. I was often paired with two or three people who were arguing completely different solutions than mine, backed by competing data. I listened to their talks again and again and read their publications. I knew where they got their data from, and I was equipped to battle them down to the mat and win the argument for my audience. My counterparts also knew what I would say and came prepared to argue against me.

I had naively believed that if I just got the facts straight and educated our world leaders and raised public awareness, there would be a global movement to address our meat crisis. Instead, I discovered that this battle wasn't about facts. It wasn't a battle of data at all. It was a battle of beliefs. Beliefs about humanity's ability to change their behavior, beliefs about the appropriate role of technology in our food system, and beliefs about humanity's relationship to animals and nature. I knew their beliefs—which acted as filters or biases, leading them to cherry-pick the facts that best supported their arguments—and they knew mine, and we were all on the global carousel of conference circuits, leaving a string of confused audiences in our wake and—devastatingly—delaying the action so desperately needed.

The reason we had competing facts was because beliefs constellate into ideologies that shape the questions researchers ask in the first place. I saw how little progress we were making, or could ever make, under such circumstances. I decided to back out of the arena entirely, at least for a while, to pursue my own research as to why we disagree so much about meat and how we might transcend our differences and learn to work together.

So, my quest began to analyze the ideologies behind the proposals presented by thought leaders on meat and sustainability. I built upon previous research done by social scientists who study competing paradigms in agri-food systems.[10] The best way I found to understand the ideologies that informed the experts was to use a method called discourse analysis to map the language used by leaders when they were giving public speeches on the topic.

Each one presented a narrative, which is essentially a story they tell. I looked for how they framed the problem, as well as who was the "we" or the hero, and who was the "they" or the villain. Finally, what was the promised land or future they would deliver us to, and how? Through systematic analysis of hundreds of public speeches by a wide range of experts in meat and sustainability, I discovered patterns in how these experts framed the problems and solutions, in what scientific data was selected to support an argument, and in how these experts related to technology as well as their beliefs in humanity's capacity to change their consumer behavior.

After clustering commonalities in language together, I found there were not just two different patterns (meat versus plants, or industrial versus organic). Nor were there dozens of different patterns. The unwieldy problem could be made legible by seeing the situation arising from four distinct paradigms, each with its own agenda for the future of meat. To simplify I have given these groups names illustrative of their approach.

The Improvers believe consumer demand necessitates increased meat production and aim to feed the world by producing more livestock in an increasingly carbon-efficient and intensified manner. The focus is on efficiency throughout the entire supply chain, from seed to feed to slaughter to plate, to scale global meat production without radical changes in our current system. Here, agricultural technologies from genetic innovations to renewable energy enable more meat with less inputs.

The Reducers believe there is no need to meet projected demand for animal-sourced foods when ten billion people could be fed by the food we already produce today—if only everyone were to swap animal protein for plant protein. Reducers are betting people are willing to either voluntarily cut or reduce animal-sourced foods from their diet or to support policies and incentives that encourage or mandate people to do so. For example: Meatless Mondays in schools or adding health insurance incentives for plant-based diets.

The Regenerators also disagree with scaling production to meet demand. But rather than aiming to curb consumption, they promote raising animals according to the natural patterns dictated by their wild ancestors to restore ecosystems, improving water storage and increasing biodiversity while sequestering carbon, and producing meat in a more humane fashion.

The Disruptors, however, envision meeting the increased demand for protein by building a future in which meat becomes entirely man-made from plants or animal cells and animals are no longer slaughtered for food. Instead of leveraging technology to squeeze more milk out of the cow, this approach utilizes biotechnology to leapfrog the use of livestock entirely.

To fairly assess what these stakeholder groups have to offer, I will walk us behind the scenes into their worlds and evaluate how each of their visions for sustainability takes into consideration human health, affordability, and animal welfare, examining them in more depth for their contributions and blind spots. While I will share the facts about the problem where there is widespread agreement as objectively as I can, my primary goal is to address the ideological reasons why we disagree about how to address the meat crisis, with the aim of helping others cut through the clutter of confusion by providing a map to navigate this wicked problem.

While I was undertaking this research, I knew I had to go beyond the bounds of academic journals and policy meetings. If I wanted to truly understand views that were different than my own, I had to go to the places where I least wanted to go, beginning with a slaughterhouse. But not just any slaughterhouse, one belonging to JBS, the largest meatpacker in the world, which globally process around one hundred thousand head of cattle, seventy thousand hogs, fourteen million birds, and sixty thousand head of lamb every single day.[11] So, before we get back to the laboratory in California with our leading-edge Disruptors, first we will go inside the system that they are trying to disrupt.

THE IMPROVERS: MORE EFFICIENT MEAT

A continuous procession of carcasses, all stamped with USDA quality grading, scrolled by in front of me as I stood in my hair net, plastic safety glasses, long white coat, and blue booties. Each eviscerated carcass, weighing around 750 pounds, had been sawed open straight down the middle. My eyes locked onto one and followed it down the line—the slick hollow of its inner rib cage and the white fat streaking throughout the body, glistening like the snow outside the factory. Just hours ago, this carcass belonged to a living, breathing, corn-munching, 1,200-pound steer. On the packing line, it will be reduced to approximately 490 pounds of boneless, trimmed, boxed beef, ready to be packed in a truck and shipped to supermarkets and restaurants across the world. This facility in Greeley, Colorado, alone processes around six thousand head of cattle every day. I kept thinking how recently that rib cage had been expanding and contracting, like waves rolling in and out of the ocean. Now, it was in an ambiguous place, a liminal space where the carcass was no longer a living animal and not yet food. It was something in between the two.

The carcasses faded into moving objects in the background scenery as I listened to our guide, Cameron Bruett, the head of sustainability for JBS.

"We're consuming about one and a half times the Earth's natural resources with our modern system, and that's not sustainable," Bruett said. "We're going to need 70 percent more food to feed all these people. At our current rate of consumption, that's about two planets, and, last time I checked, we only have one. So, there is a true sustainability challenge before us: How do we do more with less? How do we create an even more efficient system?"[12]

Gains in efficiency could potentially constitute a big step toward mitigating climate change. The United Nations estimated that if all the world's meat producers adopted the best practices and technologies already deployed by the 10 percent of producers with the lowest emission intensity, greenhouse gases in the livestock industry could be reduced by 30 percent. Intensive production systems already make up three-quarters of the world's meat (including beef, pork, and poultry), and CAFOs are predicted to make up 75 percent of the growth in production by 2030. The industry-wide adoption of the most efficient practices would represent a win-win for meat producers as well, as increased efficiency, of course, improves the bottom line. To translate those statistics: we are talking about millions of small livestock farmers around the world being consolidated and supersized into American-style mega-farms, only now it's in the name of climate change.

Bruett was straightforward and no-nonsense as he gave an unvarnished vision of what it's going to take to feed the world. Whether I agreed with this approach or not, I was relieved to know that the meat industry even had a sustainability agenda. After years of fighting against this system, I needed to understand why it is winning. This was what I wanted to know and what I needed to see. I couldn't let myself look away.

We followed the carcasses as they moved along the meat-hook conveyor belts down to the next stage in this wholesale transformation from animal to food: the packing line, where 850 workers sliced, ripped, and cleaved, separating flesh from bone and tendon from joint. As I entered the next room, I felt many of the workers' eyes skillfully lift from their tasks to look us up and down, from bonnet to booties, without missing a beat.

Efficiency means the line must keep moving as quickly as possible. Although I did not collect data for JBS or the Greeley plant specifically, data from OSHA summarized in a 2019 Human Rights Watch (HRW) report shows that in the United States, a worker in the meat and poultry industry lost a body part or was sent to the hospital for in-patient treatment about every other day between 2015 and 2018. Between 2013 and 2017, eight workers died, on average, each year because of an incident in their plant.[13] HRW found "extremely difficult working conditions, including instances where workers said they were pushed to work past their physical and mental limits" while having difficulty accessing health care. Some workers reported receiving abusive pressure from their supervisors to keep the line moving, including refusal to let them use the restroom during their shift, with many resorting to diapers as a result. Meanwhile, wages for workers in the meat and poultry industry are 44 percent below the national average for manufacturing work. Why would anyone choose to work under these conditions?

The workers were draped with chain-mail smocks. Their white under-aprons were painted with a light pink smear—almost an artistic misting—of blood. Their metallic gloved hands clutched specialized knives and hooks. Their eyes searched me. As we looked back on them, I took in faces and features from around the world and remembered the dozens of flags that festooned the hallway at the entrance to the building—the flags of the workers' home countries. The American meatpacking industry has a long history of employing immigrant labor. According to data from the 2000 Census, roughly 30 percent of workers in the industry are thought to be foreign-born, although many experts dismiss this as a low-ball estimate, since Census data doesn't take undocumented workers into account. Many employees of the meatpacking industry are refugees who speak little or no English. I couldn't begin to imagine what they'd been through on the journeys that brought them to that disassembly line in Greeley, and I wondered where they would go, if not there.

Not long after the tour of the factory, to further understand the Improvers' approach to sustainability, I attended the Global Roundtable for Sustainable Beef in a conference center near Denver. When we broke for lunch—a buffet consisting of a "make your own sandwich" bar—I shuffled down the line alongside representatives from meat companies including Cargill, Tyson, and Genentech, as well as from the National Cattlemen's Beef Association, Colorado State University, and University of California, Davis. There was also a food systems program manager from the nonprofit World Wildlife Fund (WWF). Unlike environmental non-governmental organizations that fight big corporations by protesting or lawsuits, WWF's theory of change is based on helping the world's largest companies improve the sustainability of their entire supply chains. In the case of food, if yield gaps could be closed for the major food and feed crops, 2.3 billion tons of food could be added to the global supply.[14]

I had learned from the various speakers at the event how we are amid an agricultural technology revolution—already over a $20 billion industry and growing. These technologies incorporate data science, microsensors, satellite monitoring, water-saving devices, solar power, and breakthroughs in genetic modification to increase yields of both animal feed and of the flesh of the animals themselves.[15]

In these systems, new technologies ensure nothing is wasted. Bruett had explained how lean finely textured beef (LFTB)—a technology smeared by activists as "pink slime"—utilized every scrap of meat from carcasses into food products such as Chicken McNuggets. After activist outrage, many restaurants and grocery stores have banned foods containing LFTB, forcing the industry to have to waste meat. My head was spinning. Could pink slime really be part of the sustainability solution?

I pulled a paper plate and studied the options spread out before me. The sandwich bar consisted of big stacks white and brown bread, yellow and white cheese squares, and circular meat slices in white, pink, and brown—presumably turkey, bologna, and roast beef, courtesy of the extra parts of cows like the ones I'd recently seen dangling from meat hooks on the conveyor belt.

With my plate full of iceberg lettuce and circles and squares of varying colors, I took a place at the table next to a man who had just presented named Vance Crowe, head of millennial engagement at Monsanto. Crowe had outlined how breakthroughs in data science are helping Monsanto innovate quickly and bring down the cost of developing new biotechnologies that will allow farmers to produce more feed and increase yields of animal foods. It was clear that his passion went far beyond the public relations sound bites he'd been trained to deliver, and indeed he mentioned that his mission had grown out of what he called an "ecomodernist" manifesto.

While we have yet to see genetically modified meat in the market, the public resistance to transgenic varieties of seeds that Monsanto has propagated has been far-reaching but has not slowed further developments across the industry in genetic modification. In fact, the first genetically modified animal, the AquAdvantage salmon—which grows to market size in as few as sixteen months compared to up to thirty-two months for conventionally farmed Atlantic salmon—is already FDA approved and in circulation. Very few people seem to have resisted or even noticed. Are people willing to accept eating genetically modified animals if it is a key to producing enough fish or meat to meet demand?

If we can unleash agricultural technology to reduce environmental impacts while scaling meat yields affordably, would the commodity sandwich meats prove the solution for sustainably feeding the world? As Crowe spoke, I sliced, diced, and tossed the ingredients on my plate into a roast beef, cheddar, and tomato salad. I stacked a bite onto the end of my fork and paused as I brought it to my lips. As I examined the thin slice of beef, the memory of the carcasses scrolled across my mind. Could I disassociate that image so that I could take this bite? Is that disconnection what we all do to enjoy the mass-produced meat we eat?

Improvers use what is called a Fordist lexicon, reflecting an industrial framing of food. The people who raise the animals are producers and meat is a product; animals are units of production and eaters are consumers.[16] Improvers are not exclusively bound to industry or corporations but are also found in domestic and foreign policy, agricultural sciences, and Silicon Valley. Improvers are at the helm of UNFAO, big foundations such as the Bill and Melinda Gates Foundation, the USDA, and government-funded land-grant universities.

The Improvers typically win policymakers' attention because they argue that it doesn't make a dent to make one cow or one ranch sustainable; the sustainability of all global meat production needs to be improved. They are not wrong. Given how massive the companies working in global meat production are, when they do step up to take bold action for sustainability, the impacts on protecting ecosystems are significant.

In 2021, JBS made a commitment to achieve zero deforestation across its global supply chain by 2035 and net-zero greenhouse gas emissions by 2040.[17] However, in 2022 JBS admitted to buying cattle from a criminal described by prosecutors as one of the biggest deforesters in Brazil.[18] When I read this, I feel lost. The big meatpackers operate high-volume, low-margin businesses that are constantly fielding recalls for contaminated meat, accusations of human rights violations, community lawsuits over local air and water pollution, activist outrage over deforestation, and animal welfare complaints. Yet they continue to lobby governments for subsidies and protections to perpetuate these high-risk businesses. Although there is no feasible way to solve the meat crisis without enlisting the Improvers, they still dominate our political economy, and people just keep buying their meat without asking questions.

Why? Because over the past century, the Improvers have produced and distributed meat at affordable prices and drastically reduced global hunger and malnutrition. For Reducers, Regenerators, and Disruptors to truly compete for the future of meat, they cannot ignore the Improvers' high rank in the economics of equitable access to protein. However, a major hidden cost of keeping meatpacking cheap is the reliance on the most desperate populations of refugees, immigrants, and the formerly incarcerated. Yet meatpackers are also seen as critical employers for the otherwise unemployable and an essential component of rural economies. On the one hand, jobs! On the other hand, at what cost?

While they are always "improving" animal welfare by adding ventilation or increasing cage sizes, the Improvers put animal welfare lowest on the ladder of priorities beneath price, safety, health, and environment. In fact, many environmental improvements to make animals more efficient protein machines have negative consequences on their welfare.

I am grateful for the industry leaders who opened their doors and invited me to learn from them. I hope I am seen as neither defending nor attacking the meat industry. My intention is to help us understand the world as it is so we can better imagine the world as it can be.

If we shape the future of meat with Improver logic, we might reduce our emissions and still suffer from ecological collapse, as well as human rights and animal welfare disasters. The most efficient sustainability scenario is likely to be a global rollout of genetically engineered, climate-friendly chicken powered by solar and funded by carbon credits. If we can't sustainably nourish humanity without turning the *Titanic* of the global meat industry, we must ask ourselves, is this the future we want? If not, why are letting the Improvers shape the future of our food system? Must we keep infinitely producing more cheap meat? Or could we just eat less?

REDUCERS: LESS MEAT, MORE PLANTS

In the summer of 2013, James Cameron, the Academy Award–winning director of *Titanic* and *Avatar*, and his wife, Suzy Amis Cameron, assembled a private meeting of executives from heavy-hitting environmental organizations such as the Environmental Defense Fund and the Nature Conservancy at their home north of Los Angeles. The Camerons had watched the 2011 documentary *Forks over Knives*, which demonstrated the lifesaving transformation of the protagonists when they adopted plant-based diets.[19] The film inspired the Camerons to instantaneously quit meat cold turkey and write a letter to everyone they knew, hoping that this film could help save the lives of so many friends and family who were suffering the crippling impacts of lifestyle diseases such as diabetes and heart disease.

Not long after their lifestyle change, they came upon an article that informed them that the emissions from livestock might be one of the largest drivers of climate change. As devoted environmentalists, they couldn't believe that they had been so blinded by their attachment to the meat and dairy in their diets that they had been inadvertently contributing to climate change while fighting it on every other front.

James gathered the attendees at a large, round table and graciously directed everyone's attention to Suzy, an accomplished model, actress, author, education pioneer, and mother of five children. Suzy proposed that a global shift to a plant-based diet could be the "single elegant solution" we have all been waiting for.[20] This is precisely why they also invited leading health experts promoting the benefits of plant-based eating to join the environmentalists at the meeting table.

The first part of the meeting included presentations from a few authors who were already advocates for the environmental and animal welfare synergies of plant-based diets. We learned a global shift to a plant-based diet would make a pretty big difference for the planet. For the average American or European, adopting a plant-based diet enables a 50 percent reduction in their diet-related greenhouse gas emissions while cutting their water footprint by a third. At scale, given that livestock themselves consume one-third of global crop calories, a shift to a plant-based diet could reduce global land use for agriculture by 75 percent, enabling us to feed as many as ten billion people.[21]

Studies also show that in order to stop global temperatures rising beyond the dangerous two degrees Celsius level of climate change, incorporating efficiency measures in livestock production is insufficient. Demand for meat and dairy must be reduced, in addition to mitigation from the transportation and energy sectors. However, currently, no significant climate or food policy agenda exists to reduce demand for livestock products.[22]

It was admittedly hard to think about eating after so many presentations on the dire state of our planet, but to keep the meeting flowing we jumped up to grab lunch from a buffet and bring our plates back to the table. The spread presented a cornucopia of options, including a rainbow spectrum of cooked and raw vegetables and salads bursting with sprouts to accompany a hearty bean-based soup. To sprinkle on top there were a bounty of seeds, nuts, fruits, avocado, and sauces.

While we ate, the health experts explained that, in addition to the planetary crisis, the state of global health was severe. Nutritional biochemist Dr. T. Colin Campbell, author of *The China Study*, shared his conclusions from his epidemiological study of the changes in dietary patterns in China, in which a generational switch from a plant-centric diet to an animal-centric diet caused an explosion in lifestyle diseases.[23]

Today, worldwide, two billion people suffer from lifestyle diseases caused by indulgence in what we consume (including excess meat, dairy, salt, sugar, processed foods, alcohol, and tobacco). Over the past few decades, we have reached a tipping point in history where now far more people die of overconsumption than do of hunger and malnutrition.

I paused as I held a spoon full of potato, bean, and corn soup up in front of me for further scrutiny. I looked around and saw everyone at the table helping themselves to heaping portions of freshly baked bread. I realized that this meal was full of carbohydrates. As the paleo and ketogenic diets were rising in popularity, promoting an emphasis on animal protein and fats, carbs had become public enemy number one. As someone who had once struggled with obesity and had worked hard to lose 130 pounds, I wondered if I should be worried about potatoes making me fat?

As if Dr. John McDougall, seated next to me at the table, was reading my mind, he whipped out a copy of his latest book, *The Starch Solution: Eat the Foods You Love, Regain Your Health, and Lose the Weight for Good!* His message was simple. "The human diet is based on starches. The more rice, corn, potatoes, sweet potatoes, and beans you eat, the trimmer and healthier you will be—and with those same food choices, you will help save the Planet Earth too."[24]

Dr. Dean Ornish shared about his plant-centric approach, which proved it could "undo heart disease."[25] His program has been adopted by the British National Health Service and gained notoriety when President Bill Clinton publicly turned his health around on the Ornish program.

Plant-based foods have historically been much more affordable than diets rich in meat and sweets, foods that were typically reserved for royalty. A global research project known as the "Blue Zones," which sought to explore the common denominators among the world's longest-living populations, identified simple, whole foods and plant-based diets as one of the key factors contributing to longevity.[26] The study's principal investigator, Dan Buettner, found that, in seven different areas of the world where a large portion of the population live to be over one hundred years old, most people have lived on modest means, consuming legumes, whole grains, vegetables, and only occasional amounts of animal foods as flavoring or for special occasions.[27] If eating mostly plants has been the dominant diet in so many cultures for thousands of years, how hard could it be to shift our diets back?

At the core of the Reducers paradigm is the belief that humans are fully capable of shifting our diets to save ourselves. They promote campaigns such as Meatless Mondays and aim to influence the large institutional procurement purchases that shape the menus for millions of people who eat in the cafeterias of their schools and workplaces. Although only 2 percent of Americans consider themselves to be vegan and 5 percent are vegetarian, 60 percent have reported intentionally eating less meat. The distinction means that, worldwide, the estimated value of the vegan market is $15.77 billion, while the plant-based food market is $35.6 billion.[28]

However, while we can point to motivational examples, achieving societal-scale dietary change is harder than it may seem. Even as 2 percent of the Western world is voluntarily reducing or eliminating meat from their diet, as societies become more affluent, they upgrade from plant proteins to animal proteins at the center of their plates. This is referred to as the nutrition transition.[29] While China is projected to be the fastest growing market for vegan products, it is also the fastest growing market for animal products.

Accelerating that trend, over the last century government subsidies designed to keep prices low while increasing demand for animal-sourced foods have created an expectation, or even entitlement, to abundant meat for all.[30] While reversing this trend at scale may be possible, food policy professor Tim Lang says that the only precedents for society-wide reduction in meat consumption have occurred during wartime, when citizens have been willing to accept meat rations to preserve the meat supply for soldiers.[31] Could we ever get to the point of perceiving the consequences of the meat crisis as requiring a wartime-level call for drastic measures such as rationing?

It is also hard to rely on people to willingly change their behavior when most people have insufficient access to healthy, plant-based options. While beans are cheaper than beef and plant-based meals can come at an affordable price, for many populations living in food deserts without grocery stores it can be hard to find plant-based proteins. Fast-food restaurants serving deep-fried meat substitutes are often no healthier than their conventional counterparts.

When it comes to health, while many people experience a tremendous boost in well-being when they switch to a whole foods plant-based diet, skeptics question if the health improvements are attributable to the absence of meat or the absence of processed foods and increase in plant fiber.[32]

As mentioned previously, plant-based diets have been shown by a variety of life-cycle assessments (LCAs, which assess the environmental impacts of all stages of a product) to be the most carbon efficient and would outperform poultry and even eggs for the best input-output efficiency ratio.[33] However, an outlier study found that when the digestibility of the amino acids in animal-based foods (beef, cheese, eggs, and pork) are assessed compared to plant-based foods (nuts, peas, tofu, and wheat), one would need to eat a lot more plant-based foods to achieve the same level of amino acid assimilation. Once LCAs were adjusted for the functional units of the protein quality, the environmental footprint of beef and dairy was halved, while that of wheat increased by 60 percent.[34]

How can this study be interpreted? Regenerators and Improvers will emphasize studies such as this one because they have already decided that meat is a critical part of the human diet, whereas Reducers have already concluded that meat is harmful in the diet and so they will point to environmental studies that demonstrate the efficiency of plants over animals. Reducers would say, Why not just eat plants and supplement with amino acids? This is a clear example of how studies are compared to one another while utilizing different methodologies and operating under different assumptions from the start.

Regenerators also argue that Reducers repeat Improvers' reductionist models of environmental thinking and lack an ecological understanding of the human, animal, and plant relationships that are an essential component of designing a food system that works with ecological cycles. When accounting for soil carbon and fertility, water quality and storage, and biodiversity, they argue that grain agriculture produced using conventional monoculture methods cannot meet all these goals. Plant foods are also not free of pesticides, herbicides, and water pollution, even if they are more efficient. Ultimately, a shift to regenerative agriculture is needed for plant proteins too.

At their most extreme, Reducers can be the worst culprits of us-versus-them ideological divisiveness. They will often blame all meat eaters and especially ranchers for the environmental crisis, while many ranchers would say that they are truly conservationists living in daily relationship with their animals and the land and that their use of growth hormones and prophylactic antibiotics is driven by the price squeezing of their buyers, the big meat packers.

Reducers' anger often comes from a place of righteous protection of wild and livestock animals. Many Reducers have deep compassion for animals and argue that they are sentient beings that should not be seen as food. Because of this, Reducers rank highest on animal welfare. While a world without animal slaughter would be kinder and more compassionate and humane than any type of animal farming, Regenerators and Improvers are quick to point out that eating a plant-based diet does not mean freedom from any animal death. They argue that slaughter can be humane, and that a global population living on monoculture plants would still result in competition with wildlife for habitat (such as palm oil driving deforestation in Indonesia) and all the small animals killed by harvesting equipment.

Back to my spoonful of soup at the plant-based roundtable. I'm not mad at a little starch, but given that my husband and four kids struggle to go a whole day without animal foods, I confront the resistance daily. So, I set out on a quest for the most sustainable and ethical meat I could find. This, surprisingly, led me far south of the equator to the rolling grasslands of Uruguay.

THE REGENERATORS:
MEAT THAT MIMICS NATURE

"If a president says you don't have to pay for your water because it's too toxic, something is wrong," Patrizia Cook, grandmother of nine children and prominent rancher in the Maldonado region of Uruguay, explained to me.[35] She pointed aggressively at the deceitful creek next to us, which had seemed so idyllic to me only moments before. It was the second day of a gathering of Latin American women ranchers in regenerative agriculture hosted by a nonprofit organization called the Savory Institute, and the whole group had taken a break from the talks happening inside the farm-house to walk across a meadow and down to a local stream.

After watching her grandchildren lose their friends to mysterious can-cers, Patrizia determined that the rise in local cancer rates is caused by toxic water quality, and she will stop at nothing to ensure Uruguay be-comes the first country to ban agricultural chemicals.

With short, cropped hair, dressed in jeans, boots, and a sweater, and with her lean stature and fiery attitude, Patrizia does not seem nearly eighty years old. Although she had been raised in Panama and Uruguay, her parents were American, and so she spoke English without an accent. She took a few steps away from the creek, squatted down, and pointed at the mud beneath our feet.

"I realized that if I wanted to save the water, I had to treat the soil. The soil was the secret."

She stood up and took a step closer to me, looking me straight in the eye. "The soil is the secret to everything." She stepped back and went on to explain: "The best way to make the soil healthy again, the easiest and cheapest—and it's not only cheap, you make money—is managing cattle the right way. So, what did I want to do? I wanted to make Uruguay as a small country as an example."

The herd of women began to meander up the hill, and so Patrizia and I followed them back to the farmhouse while she continued to explain to me how beef would save her country from cancer, clean the water, fight climate change, and boost the economy. We walked up a trail wide enough for the two of us flanked by fronds of Uruguayan Pampas grasses, their tops like giant blond bird feathers sweeping the sky. Patrizia stopped in her tracks and grabbed the stalk of one of them.

"Natural grasses are trees upside down. That's how deep their roots are. Uruguay is a country with over four hundred types of natural grasses in one square yard. And what's the best way to make grass grow? Managing cattle, the right way." Patrizia let go of the stalk and continued walking up the trail.

Patrizia and her cohort believe the right way is holistic management. According to the website of the Savory Institute, our host and the leading organization in the field, "Holistic Management gives us the power to regenerate grasslands from an ecological, economic, and social perspective, while regenerating Earth's desertifying global grasslands."

These women believe they can more than double beef production while reducing greenhouse gas emissions and regenerating the land. Unlike in the rest of the world, almost all the beef in Uruguay is and always has been grass fed and raised without hormones or prophylactic antibiotics, both of which have been banned for decades.

But how can this be? According to Improver logic, such measures would reduce the efficiency of production and should have put the Uruguayan ranchers at a disadvantage in the global meat market—yet Uruguay is the eighth largest exporter of beef in the world. So how is it possible to increase grass-fed meat production while reducing emissions?

Ruminant digestion of grass generates more methane than grain-based feed, and methane is the second most abundant anthropogenic greenhouse gas after carbon dioxide and more than twenty-five times as potent.[36]

Daniela Ibarra-Howell, the Argentina-born executive director of the Savory Institute, explained how holistic management sequesters carbon in the soil. "You have the bison of the grasslands in the US, you have the wildebeests in Africa, you have the reindeer in Siberia or in Alaska. They are herds of herbivores, and their whole function was to process through the microbes in their guts in their rumen. When those animals were moving through the grasslands looking for fresh grass, herbivores would never eat where their feces are. So, they will continue to move. And they move in ways to protect the herd from packs of predators, so they stay a little bit more densely packed and bunched together."[37] She explained how wild herbivores pruned the grasses as they moved, which would stimulate the growth of the roots to dig down deep and hold that soil together. Then those roots would build more soil, which helps hold the water.

She went on to say: "So, when you look at energy flow—how much of the sun, which is our source of a beautiful gift of energy that is free, right? The more plant leaf area you have available to capture that, the more you're turning that energy into carbohydrates that feed not only the plants but also the microbes, the animals, and the whole web of life." Conversely, she explained, domesticated livestock, fenced in and protected from predators, have been trained over a few thousand years or so to stay in their paddocks and rip the seeded grass up from its roots, which means "those plants start being incapacitated to photosynthesize. Then you're just capturing less of that energy, then plants die, and you start having bare ground. That then impacts the water cycle. When rain falls, most of it evaporates or runs off because the soil is unable to absorb it and retain it. So, there's less water to create and sustain all that life that you want."

Enchanted by the magical cycles of nature, when Patrizia and I emerged from the tunnel of reeds back up onto the meadow, I looked around at the grasslands stretching for miles in every direction peppered with the brown backs of cattle. But I also saw fences marking property lines. Along those lines, the neighbors have taken to farming monoculture eucalyptus trees, which I remember as invasive, water-guzzling fire hazards from my days living in California. Given the skilled labor required for holistic management—while the demand for eucalyptus is high and the labor required is low—growing eucalyptus seems an easier business than chasing cows around in circles trying to pretend that they are bison.

Back at the house, it was time for dinner. When we sat down at the table, the women passed abundant platters of vegetables and of course all types of meat they had raised themselves around the table. The devil's advocate in me dared to ask the ladies: If the goal was solving for ecological regeneration, why not just rip up the fences, put the wild herbivores back, let them do their thing, and have some beans for dinner instead? Knowing these women were skilled in both birthing and castrating their calves, I was careful with how I asked the question.

The women patiently explained to me that bean- and grain-based diets require a wide land area of row-crop agriculture and irrigation. Fields must be surrounded with fences to keep the herbivores from eating the crops, which prevents them from eating weeds, which then requires herbicides and prevents them from naturally fertilizing the soil. So, fertilization must be brought in, which is primarily synthetic nowadays, which causes nitrous oxide to off-gas into the atmosphere. Nitrous oxide is three hundred times stronger at trapping heat than carbon dioxide and takes 114 years to break down. Nitrogen fertilizer also leaches into the groundwater, making its way back to the freshwater creeks—and so we are back to giving the children cancer again.

They explained that there is a role for legumes, which draw nitrogen down from the atmosphere and fix it into the soil, as well as grains and vegetables as part of a regenerative system. But when you try to feed the world beans and grains, you must dedicate too much land to monoculture crops and trade off ecosystems.

Uruguay has the highest per capita beef consumption in the world.[38] This statistic was demonstrated for me when the women served themselves generous portions of meat. They didn't understand why anyone would choose beans over beef. They explained that beef was the most nutrient dense and beneficial for health when grass fed and free of growth hormones, and that it was all that estrogen in soy in vegetarian diets that one really ought to worry about.

Patrizia envisions a future in which the headlines for Uruguay read, "Cancer Down, Tourism Up!" "That's what I want."

Regenerators' core goal is to restore the relationship between land, animals, and humans, and they use language that frames pasture animals as heroes of ecosystem regeneration. They are challenged with building their scientific case for a different paradigm of holistic accounting in a world that has already set up systems to reward reductions in carbon emissions.

The Savory Institute has built a system to quantify and certify land regeneration called the Ecological Outcome Verification (EOV). According to the institute, the EOV "assesses key indicators of the effectiveness and health of ecosystem processes—criteria such as soil health, biodiversity and ecosystem function including the water cycle, mineral cycle, energy flow and community dynamics."[39]

An LCA of regenerative beef produced on White Oak Pastures (WOP) farms in Georgia showed that this beef has a net carbon footprint 111 percent lower than beef from the conventional US system. However (and this is a *big* caveat), the report states the assessment of enteric emissions and long-term carbon storage is highly uncertain.[40]

What does this mean? It means the more grass cows eat, the more they belch, the more methane they produce. But Regenerators insist that the Earth has always been populated with large, methane-emitting creatures and that properly managed grasslands will store more carbon than the cows can burp. But even WOP's own report admits we don't yet really have the comprehensive data to know how much carbon can be stored in the soil or for how long. This is the linchpin of uncertainty upon which the climate arguments on all sides hang.

Because of divergent ideologies, Reducers argue this uncertainty means the global warming impacts of methane probably far outweigh the carbon storage. However, Regenerators argue the uncertainty means that soil carbon storage is probably far more significant than we think.

The UNFAO estimates grassland sequestration could offset 0.6 gigatons of CO_2 equivalence per year. However, Allan Savory claims that climate change can be entirely reversed through holistic management. According to Project Drawdown, "Managed grazing can sequester 13.72–20.92 gigatons of carbon dioxide by 2050. However, this does not reduce the 10 gigatons of methane that are emitted on that grazing land today. To achieve this level of sequestration, adoption of managed grazing practices would need to increase from 71.6 million hectares to 502.1–749.02 million hectares over 30 years."[41] How could this be achieved without further deforestation? We don't know.

Since Improvers control global commodity production and influence policy, perhaps they could be enlisted to convert that 33 percent of all agricultural land that is dedicated to livestock feed crops into holistically managed grasslands?

Another challenge for Regenerators is that Improvers always sell cheaper meat. However, arguably we should pay higher prices for the additional labor required to manage and process animals holistically, the better quality of life for animals with longer lifespans, and the ecological care ranchers give to the land. But if Regenerators are to play a major role in the future of meat, there remains the challenge of making their meat available to more people. Even where I live in Boulder, Colorado—a national destination for farm-to-table dining—it is hard to reliably find regeneratively sourced meats.

Corporate supporters of Savory's work, such as General Mills and Applegate, are working to solve that. However, more consumers have bought the Regenerators' values and vision than are able to buy the actual meat itself.

Unless the Regenerators are able to solve for accessibility—or governments drastically change to stop bolstering the Improvers' models—Improvers will continue to dominate the global market, selling CAFO meat at the most affordable prices.

However, unlike the Reducers, Improvers, and Disruptors, none of the Regenerators I assessed addressed scaling their production to achieve equitable access worldwide. One unspoken and unpopular reason why Regenerators don't really model out holistic management at scale comes down to the sensitive topic of population. Media mogul Ted Turner, the largest landowner in the United States who has dedicated his land to free-roaming bison, has been quoted disclosing that the ideal number of people for the planet would be about two billion.[42] So that would leave us with a dilemma of what to do with the other six billion humans who are here now.

Reducers and Disruptors will also point out that there is a limit to the ethics of humane meat, because in the end you are still raising an animal only to kill it. Slaughter is a messy, bloody job for whoever has to do it. If we are reaching a time in history where we can meet all our nutritional needs with slaughter-free options for protein that are becoming more delicious every day, why would we continue to kill animals?

When a platter of thinly sliced beef in a nest of aromatic rosemary arrived in my hands, the entire table of women had their eyes on me. I took my fork and speared one well-cooked end piece from the edge of the platter and held it up to my mouth. I noticed the complexity of the texture and took in the aroma. Just like fine wine has the taste of the terroir in which it is grown, so does grass-fed meat taste like its land of origin. Is this something that could ever be replicated by technological analogues?

DISRUPTORS: MAN-MADE MEAT

Let's go back to the lab where the Upside Foods team was still waiting for me to try the bite of cultivated duck meat. Even back in 2017 with an early prototype, the texture and aroma were nearly identical to conventional poultry. That morsel of cultivated meat—free of all the weight of earthly trade-offs—was rare. It was precious. I knew that I had to take that bite carefully.

Then it slowly dawned on me that this pause was longer than mere appreciation. Even with all the salvation promised by man-made meat, I was not entirely sure that I wanted to eat it.

What brought me to Upside Foods' private laboratory was researching whether our culture at large could ever embrace meat grown from animal cells as a true substitute for meat derived from livestock. Since it's so hard to get people to trade beef for beans, and not everyone can afford or access grass-fed meat, I wanted to know if there was a role for the Disruptors to provide all the meat our hearts desire at affordable prices, without slaughter or environmental degradation.

Although the advances of Upside Foods have brought new media attention to cultivated meat, the idea of artificial meat is, in fact, not new at all. The fantasy of synthesizing "real" meat without the slaughter of an animal has been envisioned since the inception of the scientific revolution. In 1820, the philosopher William Godwin believed that man would evolve away from eating animals and that technology would enable replacements for animal foods. He wrote: "The food that nourishes us, is composed of certain elements; and wherever these elements can be found, human art will hereafter discover the power of reducing them into a state capable of affording corporeal sustenance."[43]

"Human art," says Godwin. The word "artificial" has become derogatory in our current era. However, reading historical literature, we can travel back to a time when the idea of making something artificial that could mimic or improve upon a product sourced from a limited natural resource was a powerfully inspiring and even romantic vision worthy of the best poetic and philosophical attention. Whether related to the environment, nutrition, food safety, famine, inflation, labor, animal suffering, or feeding astronauts, the art and science of artificial meat has been proposed during every meat-oriented crisis since the eighteenth century. Chemists have proposed that synthetic meat could be grown from algae, yeast, plankton, coal, petroleum, or even the air.[44] Perhaps the most famous quote is that of Winston Churchill, who wrote in 1931: "Fifty years hence, we shall escape the absurdity of growing a whole chicken in order to eat the breast or wing by growing these parts separately under a suitable medium."[45]

However, cell-based meat wasn't properly introduced to society until 2013 in a TV studio in London. Professor Mark Post's $300,000 cultivated patty was delicately lifted from its petri dish by a chef, slid into a pan where it sizzled as it caramelized, and served to food critics in front of a live audience. Until that moment, it was difficult to conceptualize what was then referred to as in vitro (Latin for "in glass") meat as a food, but a burger is clearly a burger. Science was now speaking our language. Burgers, we understand. Burgers, we crave.

The team that produced the burger asserted that "one single sample could produce 20,000 tons of cultured beef, enough to make more than 175 million quarter-pound patties. This many patties would otherwise require meat from more than 440,000 cows."[46] Early cell-based meat researchers commented that "back-of-the-envelope calculations suggest that a single parent cell with a Hayflick limit [the total number of cell divisions possible before divisions stop] of 75 could theoretically satisfy the current annual global demand for meat."[47] Thus, that little pink patty invited us to rethink the boundaries of meat entirely. Still, the world was waiting for someone to take a risk and take this meat out of the glass and into the world.

While Dr. Uma Valeti was leading a successful cardiology practice in Minnesota, he, along with his wife Mrunalini, decided to remove dairy from their family's diet and practice a vegan lifestyle. This was not the easiest undertaking in a small town in Minnesota, but they were committed. Many of his heart patients, however, were less committed to changing their diets, even when it was a matter of life or death.

Valeti, always keeping an eye on new research that might help his patients, began to track progress on tissue engineering for artificial organs. He wondered if the same technology could be used to grow meat for human consumption. If we could simply make meat without raising and slaughtering an animal, why wouldn't we?

While I was initially a publicly outspoken skeptic of futuristic, silver-bullet solutions like lab meat when I was out on my global road show, just like me, attitudes are changing fast. Today, major influencers in society seem to be betting that people will eat cultivated meat. In 2022, Upside Foods became the first cell-based meat company to achieve FDA approval. Today, the cultivated meat and seafood landscape includes over one hundred companies spread across twenty-five countries that have collectively raised over $2 billion. Barclays expects $450 billion in annual sales of cultivated meat by 2040, accounting for 20 percent of the meat market that year and increasing to a 40 percent market share by 2050.[48]

JBS—the very same company operating the slaughterhouse in Greeley—has recently announced a $100 million infusion into cultivated meat that includes acquisition of BioTech Foods and the construction of a new production facility in Spain, which is expected to reach commercial production by mid-2024. They are also building Brazil's first cultivated protein research and development center.

I found Disruptors to be advocates who presented livestock as an antiquated technology for producing meat. Their solution is to meet the demand by bypassing the animal and producing protein directly. Valeti called this "the second domestication," framing the "first domestication" as "wild animals to livestock." He argued, "Now we're looking at a paradigm change. We're looking at domesticating cells to grow our own food."[49]

It is important to clarify that, although cultivated meat and plant- and fungi-based meats are produced differently, they share an overall narrative and worldview that the next evolution in meat production is man-made meat. The Disruptors strive to directly compete with livestock by creating products that replace meat from slaughtered animals. They all must overcome similar consumer-acceptance challenges and policy battles regarding food biotechnology regulation and labeling. Man-made meat is presented as an innovation that would fundamentally disrupt the live-stock industry. As Pat Brown, founder of plant-meat company Impossible Foods, stated, "I know it sounds insane to replace a deeply entrenched, trillion-dollar-a-year global industry that's been a part of human culture since the dawn of human civilization, but it has to be done."[50]

Plant-based meats have already won a large share of consumer accep-tance and are predicted to become a nearly $25 billion market by 2030.[51] In 2019, the company Beyond Meat, which claims to "produce real meat, from plants" had its public offering, and within two days its value climbed from $1.5 billion to nearly $4 billion, then rallied to nearly $10 billion in a matter of weeks.

Disruptors contend that consumers are not to blame and cannot be expected to change their habits. According to Ethan Brown (no relation to Pat Brown), founder of Beyond Meat, "We love meat. We're going to continue to consume meat. There is no chance that as a race we're going to stop eating meat."[52] Beyond Meat and Impossible Foods have now led the way for an entirely new generation of plant-based meat, seafood, dairy, and egg replacements. However, definitions for alternatives to animal-sourced meats no longer fall easily into plant-based or animal-based. For example, how should we classify or label meat-like alternatives brewed from mycelium, the protein-rich fungal root systems of mushrooms that are neither plant nor animal? Or protein that has been sequestered from CO_2 drawn down from the air?

Cultivated-meat makers, however, insist on distinguishing themselves from plant and other fake meats because, they insist, cell-based meat is the "real" meat. For example, if you DNA test it, it's bioidentical to meat. What ingredients does it contain? Just meat.

Disruptors win on welfare in terms of giving people the experience of eating meat that they want without slaughtering animals. However, many animal activists continue to press for transparency on animal cell sourcing. In what conditions are these source animals living? How are they raised? Are the growth serums also free of animal ingredients? But overall, animal-rights advocates including PETA are holding out hope that the Disruptors can be the answer to easing the suffering for billions of animals a year.

The environmental benefits of plant-based meat are favorable by most assessments. Beyond Meat commissioned a peer-reviewed LCA that compared the environmental impact of the original Beyond Burger to a quarter-pound, conventionally raised US beef burger. They found that producing a Beyond Burger uses 99 percent less water, 46 percent less energy, and 93 percent less land and generated 90 percent less greenhouse gas emissions.[53]

I say *most* assessments, because the LCA of White Oak Pastures claimed that when also accounting for net carbon (including sequestration of holistic grazing), one would have to eat almost exactly one pound of WOP grass-fed beef to offset the carbon emitted from eating a pound of Beyond Meat made from commodity crops. WOP claims on their website that managed grazing by ruminants changes the paradigm from "Plants vs. Meat" to "Industrial vs. Regenerative." However, we must remember when evaluating these promotional claims that the jury is still out on methane versus carbon emissions and the capacity and term length of soil carbon storage.

When it comes to cell-based meat, researchers at the University of Oxford, also employing an LCA approach, estimated cultured meat could have up to 45 percent lower energy use, 99 percent lower land use, and 96 percent lower water use while generating 96 percent less greenhouse gas emissions when compared to conventionally produced European meat, depending on the type of meat.[54]

However, it is important to keep in mind that the supply chains of hundreds of new companies are still being formed, so there could be wide variations in the ethical as well as environmental footprints of manufacturing processes. Of critical importance is what energy sources are used to power the bioreactors. Upside Foods utilizes renewable energy in their production, but will all the new man-made meat producers follow suit?

In terms of health, both plant-based and cell-based meats are much healthier in terms of food safety and foodborne illnesses. Nearly one in ten people in the world (estimated six hundred million) fall ill after eating contaminated food, and 420,000 die every year. The primary culprit is animal-sourced foods or bacteria from animal manure on vegetables.[55]

When it comes to the nutrition of the cell-based meat, it depends on what nutrients and amino acids are cultivated and how much fat is woven into the meat. For many of the new plant-based meats, health was strategically de-prioritized in favor of taste, and thus these products don't align with the win-win the Reducers envisioned for health and the planet.

Reducers argue we shouldn't wait for technology to save us when we have an abundance of plant foods available to us now. Regenerators argue that fake meat further separates us from nature and that this technological utopianism is a distraction from the real issues here on the ground, where over a billion people in the world depend on livestock for their livelihood and play a critical role in bioregional food security. Worldwide interdependence on livestock has been a way of life for thousands of years. Many farmers are still integrating their livestock into crop production. Without animal manure, farmers would need to purchase synthetic fertilizer, with increased nitrogen runoff. Furthermore, many pastoral communities can graze their animals on marginal land, which, due to the incline, terrain, or altitude, is not suitable for any other use. Thus, the agenda Disruptors promote to make livestock obsolete could essentially also make pastoral people and their foodways obsolete.

Depending on the kind of meat, plant meats are currently two to four times as expensive as meat.[56] Cell-based meat will also likely initially be sold at a premium. All types of man-made meat aim to be price competitive with commodity meat, but they initially face similar challenges to regenerative meat of scaling a global supply chain in a way that brings down premium pricing, as well as gaining trust and establishing differentiation in the marketplace for consumers.

Partnerships with Improvers who truly understand building supply chains and mass production and distribution can help Disruptors accomplish scale. In fact, big food companies like Tyson, Cargill, Perdue, and General Mills are already investing in or acquiring plant-based protein companies. Ideally, the Disruptors give the Improvers a pathway to evolve their production model and phase out the high-risk business of mega-slaughter.

Imagine all those workers on the packing line in Greeley without blood on their aprons or chain-mail gloves grasping blades. Instead, they are packing plant-based or cultivated meat into boxes to ship around the world. This is a better quality of life for workers and better for businesses that depend on the high risks of intensive animal farming. But only if enough people will eat it.

Will these new foods appeal only to an upper echelon of society, or will a larger population really want to eat them? A 2022 meta-analysis of research across several countries reviewing factors that influence consumer acceptance of cultured meat identified neophobia (the fear of new foods) and uncertainties about safety and health as key barriers to adopting cultivated meat.[57] This leads me to believe that a subset of the population may be ready, but it may take longer than we think to bring the rest of the world along for the ride. It's not likely that animal farming will disappear in the lifetime of the readers of this book, but it is likely that there will be a lot more choices for consumers: many meats, all competing on taste, texture, price, and health, as well as their ethical properties.

Neophobia describes the adrenaline coursing through my veins during the excruciatingly long pause while the cell-based duck meat sat on the end of my fork. I'll admit I was nervous. By that point, I had thought way too much about all of this. I had worked so diligently to maintain objectivity, to question my beliefs and biases, to read competing information, and to override my own personal resistance to meat and technology. But this was the moment of truth. Would I actually be willing to eat it? A lot was riding on this. People had given up their careers to pioneer an industry that barely existed with a product that they didn't even know if people would be willing to eat.

I took a deep breath, placed the whole bite into my mouth at once, and chewed slowly, meditating on the texture of the flesh as it separated between my teeth and the juices as they slid toward my cheeks and down my throat. Without a doubt, the meat I was eating tasted exactly like duck.

I'll confess...I loved it. In fact, I enjoyed it more than meat because I had seen the whole process and didn't have to dissociate in order to eat it. I could be fully present with the pleasure. While we must assess potential negative health and safety questions of new meats with skepticism and rigor, we eat commodity meat every day knowing that we risk food poisoning and knowing it causes harm to workers, animals, and the planet. But now we also know there are several other pathways for us to nourish humanity sustainably. We now have choices.

These four bites—cell-based duck, commodity cold cuts, plant-based bean soup, and regenerative beef—embody seemingly competing agendas for meat and its role in a sustainable world. Now that we have walked through each of these paradigms and evaluated their important contributions and blind spots, we are more prepared with tools to cut through the clutter of competing claims in the divisive meat debates—not only among experts but also on menus, on packaging, or at the dinner table between friends and family.

Developing the ability to notice these worldviews is like having a magical decoder ring that makes the implicit explicit. It makes the subjective objective—not just the views of others but especially our own beliefs. By opening myself to critically reflect on my own ideology and to being willing to learn from those very different from me, I have been welcomed by the graciousness of others. I have walked between worlds, and by doing so have been able to find bridges where others saw walls.

I now believe that we can truly transcend the divisive meat debates by doing the following. First, I have learned that the deepest work of an activist is to identify from where one's own views, beliefs, and evidence originate, then to identify those of others and to spend one's energy solving for the nexus of shared values. We can start by focusing on the many areas where we all agree. Second, we must all show our cards and admit that everyone is cherry-picking statistics and shaping narratives to support our agendas. But we must also concede that each of these groups has worthy goals and unique contributions to solving the meat crisis.

Improvers' contribution is leveraging economies of scale to provide equitable access to protein, which profitably reduces hunger and malnutrition. Reducers have important dietary behavior change interventions to reduce lifestyle diseases while simultaneously reducing animal suffering. Regenerators have a deep understanding of ecology to help heal nature while producing more humane meat. Disruptors have the biotechnological tools to separate meat production from its historic extractive dependence on animals and nature while giving consumers the meat they desire. From here, we can cocreate the future of meat.

So many people are resigned to the world as it is, and that disables them from participating in creating the world that could be. We fail ourselves when we believe that the system will never change, because all our food system has ever done is change. Food is the underlying foundation of society itself, and animal-sourced foods are the most prized and central component in most cultures. Over millions of years, our food system has evolved from hunter-gatherer to horticultural to agricultural to colonial to industrial. As it evolved, our technology for growing and processing food has become more complex and sophisticated.

For me, understanding that agriculture is only ten thousand years old and that the factory model for animal farming is less than one hundred years old made me realize that those of us alive today who truly dedicate ourselves to the cause can absolutely change our food system. In fact, it's already changing all the time. But according to who's agendas and with which values and what goals?

What if we designed the future of meat utilizing the best of what each of these stakeholder groups has to offer? To do so, we must also remember that each agenda has blind spots, and when one dominates, those blind spots become an externalized shadow over our world. The harm of one can often be remedied by the unique medicine offered by another—if we can be willing to admit our own blind spots and work together.

When it comes to our own dinner plates, we are tribally wired to be suspicious of people who eat differently and to have more trust in those who eat like us. Although I prefer beans over beef (I have been known to endearingly call my bowl of beans "little pockets of magic"), when choosing for my family, I seek out meat providers that are actively working to regenerate their lands or sustainably harvest wild meats. I am grateful we have the privilege of choice and recognize that so many do not. But when I am a guest of others in their homes, communities, farms, restaurants, conferences, or laboratories, I eat whatever I am served with gratitude.

That is precisely why I ate each of the four bites offered to me in this chapter, so that I could genuinely connect with and understand the world-views of my hosts. But if I defined my identity by my food choices, it would divide me from other tribes, including my own family, and I will not let my personal food choices prevent me from collaborating with someone who eats differently but who wants to work toward designing and building a better world.

Without expertise to offer on competing health claims, I will only suggest that humans are biologically omnivorous beings that have lived on widely varying diets across the planet for millions of years, so it is vital to remember that we are remarkably flexible. While there is a lot of compelling new dietary research, our sciences are still relatively young in trying to understand what optimal health for humanity really is—or can be.

However, I have found in my research that ethical claims often become conflated with dietary advice and vice versa, so it's important to remember that just because many people can be healthy eating exclusively plant-based diets doesn't mean these diets are nutritionally, culturally, or ecologically viable for everyone on the planet. Just because our land can be regenerated by integrating livestock animals doesn't mean we should all eat lots of meat. Just because meat contains important nutrients and is a beloved component of most cultural cuisines doesn't mean we should justify and dissociate from human and animal suffering and environmental degradation.

Charles Eisenstein writes, "When you eat something, you eat everything that happened to make that food come into existence. You are affirming a certain version of the world." He then asks, "Does it nourish you? Are you happy with the reality you are saying yes to?"[58]

Our job here and now is to reconnect to our food so we can be more fully present with the pleasure of each bite. If connecting to the journey our food has taken to our fork reduces our pleasure, we should reconsider what we are eating.

With nature on our plate three times a day, it's important to understand the systems that shape each bite and their impacts on people, animals, and the environment; to identify our own values; and to eat in a way that is in alignment with those values whenever we can. Simultaneously, we must have compassion for ourselves and others by acknowledging that it's nearly impossible for most people to have a perfectly ethical diet when choosing from foods produced by a broken food system that is not fair, healthy, or sustainable.

We also must reach beyond the boundaries of our own plates and find alignment with unlikely allies to mobilize a better future for meat. We must not disconnect. We must see the world as it is—with all its complexity, all its unintended consequences that we sweep under the rug in the name of progress. This is our world that we share, and we are all impacted by ripple effects of a broken food system. There is a role for absolutely everyone in shaping a better future for all of us. Together let's paint a more beautiful world in which people, animals, and the planet are thriving.[59] We must first believe it is possible and then fight to bring it into being. But whatever we do, we must do it now—not only to avoid the worst but also to lay the foundations for a flourishing future. It's not too late.

15.

From Food Services to Foodshots: Notes from an Unexpected Change Leader at Work

By Michiel Bakker

Michiel Bakker is an accomplished senior strategic, multidiscipline global services and experiences business leader, grounded in hospitality with a touch of technology. As the visionary behind Google's celebrated flagship Food@Work program, he brings deep experience in global operations, people, project, program, design and development, and change management.

He leads Google's Real Estate and Workplace Services (REWS) Global Programs Team, which manages programs related to food, transportation, sustainability, global events, amenities, and guest services, as well as health and performance. He has overall responsibilities for the design, development, and ongoing delivery of REWS global workplace programs and services to over 250,000 individuals in more than fifty-five countries around the world.

Before joining Google in 2012, Bakker spent two decades in global hotel and food-and-beverage operations, creating and building out partnerships and leading hotel food-and-beverage developments, openings, and food experience design.

CHICKENS AT GOOGLE

It's September 2012, and I'm standing in the parking lot of the Google campus in Mountain View, California. Chickens are running around in a temporary henhouse. A bus pulls up. The door opens, and then it hits me. The opportunity to act on my belief and desire to contribute to creating a better food world is real. It is actually *happening*.

That morning, the launch of what was to become the Google Food Lab was the beginning of my unexpected journey through the world of food and Google—and, along the way, embracing the notion that, as individuals, we can use the personal and organizational resources at our disposal to make a positive difference in society, if we choose to do so.

For better or worse, I had seeded my future in food. I had no idea, however, how hard, messy, frustrating, exhilarating, and ultimately rewarding my journey was going to be.

My story is a living example of how you can use your role at work—or any role at any organization, for that matter—to become an agent of positive change. While I am the first to acknowledge that I work in an unusual organization with plenty of resources, I believe my experiences are relevant to almost anyone who is seeking to be an agent of change.

Like so many people, I try to fulfill multiple identities and roles in my life. As a son, a father, a husband, and a leader at work, I have hopes, dreams, beliefs, and flaws, as we all do. Consciously or subconsciously, I want to make a difference in this world. It is not that I feel the urge to protest every day or address every wrong, but I want to take care of my part, my contribution to the communities I belong to. And based on my desire to actually make a difference on the complex challenges we face, I have learned firsthand over the years about the ability to believe in oneself, to act on that belief, and to keep going, even when we're out of our comfort zones. I have found ways to become a little bit more comfortable in challenging situations. I've also learned how this ability to tolerate being uncomfortable, not knowing the answers yet continuing to nurture belief in oneself, is actually at the heart of what it means to be an effective changemaker.

In today's food worlds and foodways, we need more individuals to step up as change leaders and changemakers. There is so much to be worked on and to be addressed that we need all the help we can get. However, there are real barriers to many of us filling these roles, and I'd like to share how I navigated this on my own journey. Regardless of whether you work in the nonprofit or private sector, in start-up land or in finance, education, or policy, no matter what discipline you practice, I invite you to come along for the ride, believe, and seed the food future(s) you want.

MY INITIATION

If you are engaged in the challenges of today's food worlds, it is easy to agree with broad statements like, "Our food systems are broken" and "They need to be fixed." How many times have you heard these statements uttered in meetings, events, reports, and grand proclamations? How often have you made them yourself? Many of us have firsthand experiences of what we believe to be wrong with our food systems, what needs to be fixed, and what we would resolve if we had a magic wand. Reduce food waste, cure food insecurity, create better working conditions for farm-workers, pay food service workers a living wage, ban plastic straws—the list goes on and on.

I was there with you with the belief that food systems are broken, so let's fix them. Working as the regional food-and-beverage leader for the company Starwood Hotels and Resorts in Europe, the Middle East, and Africa in 2010 and 2011, I was coming of age in my understanding of food systems and my food beliefs—that is, my various ideals and values relating to food such as culture, nutrition, security, equity, and joy. I was starting to see the broader, more systemic needs, opportunities, and challenges in food systems. I even dipped my toes in the water for a bit by spending time with Starwood's food-and-beverage leadership cohort on the development and launch of our first-ever regional sustainable seafood policy. Was that policy helpful? Did it make a difference? Most likely, not enough. But at least we made a start and got going. That experience got me thinking more about our theories of change, especially in the business sector. How do you actually affect change or make an impact?

People who know me well recognize that I am a curious individual who loves books and frameworks. I find inspiration and get energy from reading across a wide range of disciplines and fields, especially about subjects adjacent to my core areas of work and interest. I enjoy noodling about insights picked up and creating frameworks and strategy maps. Just ask my food team how many versions of a strategy map for the food program you can create. (Hint: it's *a lot*.)

During my hotel days in Europe, I encountered a framework that actually provoked me and evolved my default ways of thinking about the business world. I picked up the book *The Responsible Business* by Carol Sanford. Its core message was that as a "responsible business," not only are you responsible for generating the required (and dare I say "traditional") returns for your shareholders, but you should actually work on finding a balance between the interests of five different business stakeholders: customers, co-creators, the Earth, the community, and investors. This was truly challenging to my professional beliefs, which I had developed over the years and been trained in at college. But gradually I understood the framework on a philosophical and conceptual level and mentally embraced it.

Of course, you might say that this concept of "conscious capitalism" or "stakeholder capitalism" wasn't new. You may even be wondering, "What took him so long to think broader than just financial returns?" But isn't that true for so many? What is also true is how much of the knowledge and insights needed to change the world are already here; they are just not equally distributed and known to all. They often stay in echo chambers, being heard by just a select few.

Thankfully, a new (to me) way of thinking had found me. I found the book inspiring, and it made me think more—and differently. What I did not get then, however, was how to use this philosophy to actually create change and make impact. In my own mind, I was just a senior middle manager at a large global company with many different stakeholders with conflicting interests and priorities. I was clearly unable to affect change, or so I thought. My unspoken private belief at the time was that you had to be in a senior leadership position to make a difference, which meant that it was not something I truly had to think about. I had basically exempted myself from seeing myself as a change agent based on assumptions and beliefs I had been carrying for years.

However, the seeds for changemaking were planted inside of me with Sanford's book and started to sprout. I had been touched by the utterly compelling yet daunting prospect of actually being able to make a meaningful, measurable impact in service of the larger whole—impact that could actually make our world better.

A year later, standing in that parking lot, I was now leading Google's global Food@Work program. Never had I envisioned during my twenty-plus years in hospitality that I would ever work for a global tech company. I was genuinely surprised when I got an email from a Google recruiter in the first half of 2011, asking whether I might be interested in joining the company. While I obviously knew of Google, I had personally not heard of its food program. I was having a lot of professional fun in the Middle East, so I responded initially that I was not really interested. But Google intrigued me, so I thought: Why not have a chat or two and learn more about the company? After all, I had nothing to lose, right?

During the subsequent conversations and interviews, the notion of working for a company where people view their core mission as making a difference in the world started to deeply resonate with me. A key moment was a conversation with Dave Radcliffe, the vice president of Google's Real Estate and Workplace Services, the parent organization for the food team at Google. Predictably, he described in glowing terms the opportunities with the role and how fabulous it would be to work at Google and for his organization. That actually wasn't what compelled me, however. He got and kept my attention when he shared his belief that those working at Google had many opportunities to actually make a difference in the broader world, including with Google's food program. How they would do this with the food program, he did not know yet, but he was looking for an individual who would be able to see and create such opportunities and, more importantly, act upon them.

Radcliffe saw something in me that I had not yet seen in myself. He saw me as someone capable of creating something new, maybe even visionary. He saw in me a change leader before I did, and that got my attention.

Now, years later, I can see that the decision to accept the job offer and join Google was my best career decision. It began the gradual process of me redefining and evolving the scope of my role and responsibilities, expanding my identity from "just" a middle manager and operational leader at the bottom of the food chain to identifying myself as a systems change-maker, leader, and enabler, as well as someone who saw. that workplace amenities could be used for more than solely serving an organization's employees. It took me years, but this process radically and fundamentally shifted how I related to the impact I could and would start to make through my role at work. Google became a working laboratory for me, as well as for countless others, for affecting change in food systems.

In short, my journey at Google taught me how you can become a changemaker at work, even if you are not the top dog.

FOOD AT GOOGLE: MORE THAN JUST FEEDING GOOGLE'S EMPLOYEES

Food has been an integral part of Google since the company's beginning in 1998, and food is taken really seriously by all in the organization. Why would a global tech company be so focused on food at work? Google's leaders deeply believe and embrace the notion that food at work enables productivity and performance, that good food supports Google's culture and work dynamics and enables Google to retain and attract healthy top talent. The program also enables casual collisions, togetherness, and creativity. Google's food program truly occupies a unique space in the world of corporate food services due to its size and global reach.

Today, Google's food program is globally recognized as aimed at fostering organizational culture, enabling people to thrive, and touching food systems and worlds beyond the company walls. Its mission is to inspire and enable the Google community to thrive through food choices and experiences in the hybrid work environment. With an amazing leadership team and the support of our cherished service partners, it has grown over the years into a global internal food services and experiences organization.

As you can imagine, the Food@Work team has been iterating on and evolving the program over the years. We have researched and experimented with getting individuals to eat their vegetables and drink more water; we introduced the design, building, and offering of teaching kitchens at work; we activated a set of bold food-loss and waste-reduction commitments; and we built tens of workplace cafés and hundreds of micro-kitchens around the world. Today, we are responsible for providing all individuals who work at Google with delicious, nutritious, and sustainable food choices through a global portfolio of cafés (restaurants), micro-kitchens, food trucks, and teaching kitchens. One can see Google's food team as the master franchisor of its food program, with the actual food services being delivered on a daily basis by a variety of world-class food service providers.

Over the years, it became clear to me that our food program's guidelines and approaches did definitively impact our employees and the broader culture at Google. However, even at a data-driven company, I found it difficult to define precisely the true impact of our food team's program decisions with clear data. Our relationship with food, the ways our lives are impacted by what we eat—emotionally, nutritionally, and even in terms of productivity—can be difficult to measure or even articulate. But the qualitative and anecdotal feedback we received was compelling, and, luckily, Googlers were not shy about giving their thoughts. We would hear from them daily through our internal platform called Foodback.

We also got quite a bit of qualitative feedback every other year with an internal Google-wide survey called Google Eats. Googlers told us that the longer they worked at Google, the healthier they ate at work and, even more importantly (and surprisingly), at home. Apparently, based on what we were hearing, consistently enjoying healthy food (with "healthy" being our default option at work) can change our palates, our relationship with food choices, and our food beliefs over time. As a result, we knew our offerings were popular, and we sensed people were healthier.

When it came to meat consumption, emotions ran higher. A Meatless Monday initiative at Google in 2010—inspired by Google's "Optimize Your Life" campaign, an internal company-wide healthy lifestyle program—won very few converts. The food team learned experientially that trying to change behavior by taking away a beloved ingredient does not work. Influencing and changing food behaviors requires a truly holistic approach, including addressing menu offerings (for example, by making the better choice the default choice) and through demand strategies (such as providing more information and narratives to enhance food literacy, as well as offerings such as cooking classes).

While we were advancing the program with significant step changes in the scale, scope, breadth, creativity, imagination, and experimentation used in serving Googlers, it started to dawn on me that we might be in a position to make a positive impact on the broader world. I initially had not given a lot of thought to the impact we could be having indirectly on the larger food industry. My focus was primarily on establishing an innovative, best-in-class food program that had the potential to dramatically impact the organizational culture. But that changed when I started to hear about food teams at other organizations asking our food service partners about what we were doing. Representatives of large food-and-beverage companies were also requesting us to share our insights. Food service associates who worked on the Google account migrated over time to different organizations, infusing food programs at other companies with Google's food philosophies, such as using food to bring people together at work, actively using behavioral economics throughout the program, and cooking from scratch as much as possible. Others were inspired and intrigued by what we were doing. People were truly paying attention. Our approaches started to ripple through the industry.

TARGETING MY SPHERES OF INFLUENCE:
THREE ECOSYSTEMS

Although it has taken several years to articulate this concept in the way I do today, I was experiencing a gradual yet radical reframe in how I understood and defined my role—and how we unintentionally curtail our actual impact. It has to do with the assumptions we make about what we can and cannot do, and the (usually) unconscious ways we limit ourselves. If I define my work role as "just" being responsible for the food program at Google, my impact will be fairly limited to what I and the team are able to do internally. Instead, imagine that I define my role and my spheres of influence much more broadly. Why could I not do that?

These days, I believe that I can and should use my role to impact three ecosystems. The first is my core, traditionally defined area of responsibility, delivering amazing food experiences to Googlers at work on an ongoing basis.

The second ecosystem is the world of Google and Alphabet at large. Impact in this ecosystem is made by creating value by actively partnering with other teams at Google and using our program as a working lab for our partners and others. The biggest impact area of contribution, the third ecosystem, is the world at large. How can we directly or, through other teams, have Google contribute to addressing global food systems challenges?

THE GOOGLE FOOD LAB: AN UNEXPECTED EXPERIMENT

As I was forming my approach to bringing these ecosystems together in those early days, I continued to learn what from my previous professional life simply would not work in this new environment. Having worked for a large global hotel company for over seventeen years, I was used to collaborating with hotel owners, asset managers, subject matter experts, franchisees, food-and-beverage leaders and associates, and suppliers on aligning on food-and-beverage initiatives in a wide variety of settings such as councils, advisory boards, committees, or task forces.

The hotel industry is an environment in which your success depends on your ability to work with other stakeholders and find common ground. A hotel franchisee (a company granted the rights to use a specific hotel brand) will usually want to maximize the benefits it can derive from the brand affiliation while at the same time maintaining maximum freedom to run the operations in the way it sees fit. On the other hand, the hotel franchisor (a company that grants another company the rights to use a specific hotel brand) is dealing with perceived tensions such as maximizing the value of the brand and the profitability of its franchisees. To address these kinds of issues, the various stakeholders periodically meet in settings like franchise food-and-beverage advisory boards or franchise councils, where new food-and-beverage standards and initiatives are collaboratively explored, developed, and agreed upon.

As an industry leader, I've learned that the process of developing something together with multiple stakeholders—with often very different perspectives, requirements, and goals—can be arduous, messy, and time-consuming. But the outcome tends to be better, more holistic, and more readily embraced and supported by all than when a single organization tries to dictate decisions for the entire stakeholder group.

I brought that depth of multi-stakeholder experience with me to the food team at Google, with the assumption it would transfer over pretty straightforwardly. I was eager to build upon the amazing Google food program foundation and take it to the next level by partnering with internal and external stakeholders. However, I quickly realized that Google worked differently than just about any other organization I had experienced before. In light of the need to infuse our work with internal and external thought leadership, while being attentive to the unique attributes of our culture, I wondered how we might engage outsiders in exploring the future of food-and-beverage experiences in a way that resonated at Google and created value for the outsiders. This was an active topic of discussion with Michelle Hatzis, a healthy lifestyle program manager at Google who had started around the time I did. We hashed this out during the summer of 2012, playing the what-if game. What if we organized a summit of partners we had worked with in our past to create food experiences for sustainable high performance within Google and beyond? We could call the gathering an "innovation lab" and see what happened.

Tuesday, September 25, 2012—the morning I stood in the parking lot in shock as chickens ran around in a temporary chicken pen and a bus filled with eager participants showed up—was the inaugural convening of Google's Innovation Lab for Food Experiences (later shortened to the Google Food Lab). Our initial cohort of fifty-plus participants included CEOs of food services companies, representatives from food retail and design firms, professors and academics in the field of health, nutrition, and public policy, farmers, food operators, and representatives of big food-and-beverage companies and foundations. It was an incredibly dynamic and rich community, many of whom had not had the opportunity to engage in a multidisciplinary gathering—a lab—where the sole purpose was to share ideas, experiment, and see what emerges.

The fact that Michelle and I were able to get that many food-focused changemakers and subject experts to come together for a couple of days in Mountain View to explore the future of food experiences made me realize that we were onto something. The initial act of bringing people together turned out to be the easiest part. During the two days, we broke bread and enjoyed some good Google food together, and each participant was able to learn a little bit about the others' challenges, worlds, remits, concerns, and cares. We heard an array of perspectives and started fostering a sense of community. But figuring out what we could jointly work on turned out to be less easy and straightforward. While many of us in the room agreed that the food systems needed to change, we disagreed about what aspects needed to change, why, where, and how. We were all over the map with our own beliefs, theories of change, and preferences.

During 2012–2014, we experimented with different themes and directions for the lab. We gathered twice a year based on our core belief that if you bring deeply passionate, committed, and action-driven change leaders together, change might happen or be accelerated. While many of us felt that there was something there—participants were inspired, and there was a palpable energy of community and synergy forming—it wasn't clear at all in the beginning what our actionable North Star was going to be. Should we, as representatives of various organizations, collectively work on hunger in Africa, improve school food in the United States, reduce food insecurity in rural areas, or dig into food-related health issues?

There were definitely moments during the initial years of the Food Lab when I felt that this was just too hard, fraught, complex, and not leading us anywhere. But Michelle and I were 100 percent committed to making the Food Lab a success for all involved. Ultimately, as I'll explain later, some valuable partnership projects emerged from the Food Lab conversations.

My takeaways from the initial Food Lab effort? One, don't give up too early or when something is frustrating. Change is hard, truly hard. Two, flip the script around. Instead of bringing change leaders together and exploring what you might want to work on, define what it is you want and find like-minded people to partner with. And three, when possible, use your position to serve as a convener, facilitator, and host for diverse, stimulating, and ideally challenging conversations that move ourselves, each other, and the movement forward.

FROM MOONSHOTS TO FOODSHOTS: ACTING LIKE AN INNOVATOR

During these initial years at Google, I quickly learned how to be (somewhat) effective within the company's complex organizational structure. As I developed my internal network, I realized that Google—like many complex, innovation organizations—runs on relationships. In 2013, I serendipitously met Emily Ma, who at the time was the head of global operations at Google Glass. Emily introduced me to the concept of "moonshots," as driven by Google X, the corporation's research and development org and the incubator and accelerator of big ideas such as self-driving car technology, balloon-based internet access, and Google Glass, which Emily herself was working on.

Moonshot is a metaphor for a large, risky challenge that requires a significant amount of effort and resources to address and resolve. You can think of a moonshot as a "big hairy audacious goal" as so eloquently defined by Jim Collins in his book *Good to Great: Why Some Companies Make the Leap and Others Don't*. Moonshot thinking is about showing up with the willingness to challenge broadly held beliefs, to think creatively, and to consider unconventional approaches to problem-solving.

Different ideas started to come together in unexpected ways in my mind. Aren't food systems challenges just as complex as the many technology challenges being explored at Google X—or even more so? And why should only tech teams be able to come up with moonshots? While the food team provides a support function and is not a Google business, nobody at Google had ever told us that we could not come up with and build out our own moonshots. We already had an audacious rallying cry: How might we contribute to enabling the planet to feed and nourish ten billion people by 2050 in sustainable, inclusive, efficient, nutritious, and healthy ways?

With my desire to make a positive impact in food systems and being in the fortunate position of leading a global Food@Work program at an organization with significant resources, filled with individuals and teams eager to make a difference, maybe I could define some moonshots for us to pursue. I began to wonder about potential moonshots in the world of food in which we could make a difference based on what we were already working on. What were some incredibly big challenges in food that could benefit from a very different approach?

I sensed that our moonshots should not be as broad and vague as "solving world hunger" or "saving babies." We wanted a big goal that was also specific and attainable. I combined that notion with Jim Collins's hedgehog concept. Based on his research, Collins advises would-be changemakers to focus on the magical intersection of what you are deeply passionate about, what you can be the best in the world at, and what drives your organization's economic engine. What could that intersection be for us?

One of the desired futures of our Food@Work program was to help individuals make better food choices: to eat more plant-forward food and to lessen the unintended consequences of our diets. But this raised the question of how you enable and support consumers in making better or more informed food choices. Beaten-path thinking said that the answer involved requiring restaurant operators to list all the ingredients and nutrition information on menus, regulating maximum portion sizes, and educating consumers about the dangers of practices like eating too much salt. But if all these traditional measures were effective, why did we continue to see an increase in diet-related diseases all over the world? I was starting to deeply question the prevailing theories of change that had so many of us enthralled, from behavioral economics, nudge theory, and information feedback—tactics pioneered in the public health sector—to behavioral science insights in choice architecture.

Combining this goal with Google's core competencies and interests (data management, insights, technology, artificial intelligence, health, and well-being), I led the development of possible broad themes and started to experiment with them in different settings and with different audiences. I observed how people responded to possible ideas and the terms used. As is so commonly done in a tech organization, I launched and iterated on the key ideas for months. Over time, the five themes became clearer and actually started to stick. You know that your ideas are having some initial success when others use them when you are not around.

And so our five Foodshots were born. We asked:

1. How might we enable individuals to make personal, informed food choices for sustainable lifestyles?
2. How might we shift individuals and populations to a more balanced, plant-forward or plant-rich diet?
3. How might we enhance food systems transparency and traceability, organizing the world's food information and making it universally accessible and useful to all?

4. How might we reduce loss and waste in food systems?
5. How might we accelerate the world's transition to a circular food economy?

These Foodshots defined five long-term aspirations that we members of the Google food team could use to guide and direct our efforts, as we were eager to make a broader impact on the food world as well as on our own food program. Truly affecting change in these areas would be immensely meaningful not only for the food team at Google and its food program users but also for our stakeholders, internal and external, and for the world at large. Imagine, for example, that we could develop new ways for individuals to make better food choices. This would be incredibly impactful not only for our own employees but also for eaters outside of Google's Food@Work ecosystem.

For me, the Foodshots were initially directional, pointing to desired future states but lacking clearly defined end goals. Foodshots don't include specific, measurable, achievable, relevant, or time-bound outcomes (unlike the objectives in the well-known SMART Goals framework). I struggled with this notion for quite a while. I come from an operations background, and I love frameworks and measurable impacts, and I still find working without them very challenging. How do you define success for a Foodshot? When can you say, "Mission accomplished"? What is the end state for a Foodshot like "shifting diets"? Will you ever be done? And how do you make trade-offs between various worthwhile Foodshot initiatives? Which initiatives will you support based on which criteria, and which ones will you pass on?

I still don't have satisfying answers for these kinds of questions, but I am more at peace with the paradoxes that arise when they remain unanswered. When you are pursuing big, hairy, audacious goals, there is always way more work to be done than you will ever have capacity for. If you are not careful, you can spend all your available energy pondering and talking about what to do or not do. For me, the five Foodshots have become a filter to help me determine what to focus and act upon and what to leave for others.

Work on the Foodshots is ever evolving, iterative, and ongoing. The concepts of moonshot thinking and Foodshots have also enabled me to seed new initiatives in the spirit of exploration, understanding and accepting that not all ideas will lead to desired end stages.

The concepts do help to initiate and guide action and learnings. For example, as part of our Foodshot to enable individuals to make personal, informed food choices for sustainable lifestyles, we partnered with the Yale School of Management to conduct research on food choice architecture. We investigated the impact of changing food product order and location, offering more intentional portions, or using descriptive dish titles to influence food choices.

To advance insights into shifting to more balanced, plant-forward diets for planetary sustainability and individual health, we provided seed funding for the Food for Climate League, a nonprofit research collaborative working to identify the optimal communications tactics and avenues for engagement to catalyze a global movement toward climate-beneficial eating. We also partnered with the Culinary Institute of America on a new plant-forward culinary arts education and certification initiative. This initiative supports organizational transformation around plant-forward menu innovation, skilled execution, engagement, and adoption, significantly contributing to larger food system change through the training and inspiration of future food leaders.

To further enhance food systems transparency and traceability—our third Foodshot—we partnered with Google X on a seafood traceability initiative. This initiative connected suppliers, buyers, and technologists in Japan to explore how technology might contribute to a sustainable future of fisheries for all, starting with increasing the amount of real-time data coming from their supply chain. This is now a core project at X, called Chorus (you can read about it at https://x.company/blog/posts/introducing-chorus/).

To learn more about how to reduce loss and waste in food systems—our fourth Foodshot—we partnered with a company called Full Cycle Bioplastics to build a demonstration prototype on-site at Google to convert all of the company's Bay Area organic waste into PHA—a naturally occurring, marine degradable plastic alternative—which we are using to prototype compostable wares that may replace traditional compostable products used throughout our Food@Work program.

These are just a few examples of the various ways we have activated ambitious, scalable approaches by leveraging our ecosystem and partnerships.

LESSONS LEARNED ABOUT DRIVING
IMPACT FOR GOOD AT SCALE

In my leadership role at Google, I've been hugely fortunate to spend a lot of time thinking about and, more importantly, working on broader food systems challenges over the years. After twenty Google Food Lab summits, ten years on the EAT forum advisory board, growing the food program from serving 35,000 people in 2012 to over 250,000 in 2023, and being an active support partner in various Food for Good initiatives, I can say it has been a fascinating journey so far. While my evolving change-leadership beliefs have often been imperfect, I have learned by doing and now have a deeper understanding of and appreciation for the messy realities of broader systems change leadership through one's role at work.

What held me back as a food system change leader for the longest was not my lack of technical expertise or experience. It was my original unconscious and deep-rooted belief that I had to do all if not most of the work myself, or at least be heavily involved in it. I never wanted to be seen as one of those leaders just bossing other people or telling them what to do; that style of leading has always been anathema to who I am and what I believe. But guess what? Over the years, I realized that believing that I somehow had to "own" or be involved in everything my team did actually held me and my change impact back. You limit your impact-making capacity if you want to be involved in all activities and initiatives, as opposed to empowering others. You literally become the bottleneck.

Letting go of this self-created notion was liberating for me. My dreams for better food systems are our five Foodshots, and I believe that is our sweet spot, our niche. But it doesn't mean that only work done or led by me or my team is relevant, impactful, or worthwhile. In fact, the more we can get everybody to work on our dreams, the more we can get accomplished, faster.

Over the years, I've fully embraced the notion that my team and I can play many different roles in leading, guiding, and enabling systems change. I can be a driver, a funder, an investor, a supporter, an endorser, a tester, a collaborator, a critic, an activator, a facilitator, an entrepreneur, an intrapreneur, and so much more, at various times or all at the same time, depending on what's needed. The same applies to you. Graciously embrace and celebrate all the roles you can play in enabling and leading the change you want to see. All roles matter.

CHOOSE TO BELIEVE YOU CAN—AND WILL—MAKE A DIFFERENCE

Perhaps my biggest learning overall is that becoming an effective systems change leader has very little, if anything, to do with your actual formal work role, your level, or the organization you work for. For me, it starts and ends with you truly and deeply believing that you can, want to, and will make a difference with the role you are in. While it isn't that easy, it is that simple. You must be willing to see the societal and systems challenges at hand and be able to envision possible contributions toward systems change you can make by stepping up and using your role, your ecosystem, and your network to contribute, no matter how big your perceived contribution might be. It does not matter whether you are in a large or small company, a for-profit or nonprofit organization. What matters is that you believe you can make a difference and your willingness to act upon that belief.

You also have to be comfortable with pushing boundaries and seeing how far you can get before the system shuts you down. When we started the Innovation Lab for Food Experiences in 2012, it was definitely a very different approach than how we worked during those days. However, I was willing to take personal and professional risk and ownership for the initiative, to ask my boss for forgiveness afterward, and to always look for synergistic opportunities. In the long run, the risk-taking paid off.

Change starts with you, but it doesn't start until you do, as executive coach Tom Ziglar reminds us. If you want to be a changemaker, whether in regard to our food systems or in some other arena, it is time for action now. Where might you get started? How would you define your ecosystems at work? What moonshots do you see emerging? What roles can you play?

Most importantly, do you believe in the impact you can make? If the answer is yes, then there's no reason not to get started now.

Part III | **WHAT YOU CAN DO**

16.

Finding Ways to Nourish Both People and the Planet

By Danielle Nierenberg

Danielle Nierenberg is a world-renowned researcher, speaker, and advocate on all issues relating to our food system and agriculture. In 2013, she cofounded Food Tank (foodtank.com) with Bernard Pollack, a 501(c)(3) nonprofit organization focused on building a global community for safe, healthy, nourished eaters. Food Tank is a global convener, thought leadership organization, and unbiased creator of original research impacting the food system.

Nierenberg has an MS in agriculture, food, and environment from Tufts University's Friedman School of Nutrition Science and Policy and spent two years volunteering for the Peace Corps in the Dominican Republic. She is the recipient of the 2020 Julia Child Award.

This book has taken us on a journey through the complexities, nuances, and unjust realities of the way we produce, distribute, and dispose of food globally.

We have a food system that focuses on filling people up—with commodity crops and ultra-processed foods—rather than nourishing them.

At the same time, we don't give the respect that is due to food workers or farmworkers. Black and Brown folks and Indigenous communities who have championed regenerative practices for generations continue to lack recognition, resources, and food sovereignty.

Thankfully, there is a bounty of solutions for our food, agriculture, and climate crises. Through my work as president of the global food nonprofit Food Tank, I am usually on stage or in front of a microphone interviewing food systems leaders and changemakers. I get to ask my personal heroes their deepest thoughts about food systems change and what it will take to transform our agricultural policies. It's a privilege to talk with so many experts from around the world and with a wide range of interests. And it has perhaps given me a unique vantage point on some of the important, overarching issues we face today.

There are countless stories of hope and success across every region and every level of the supply chain. These are the stories that aren't covered often enough in the mainstream media. While the stories of inequities, abuse, and corruption must be shared, the stories of progress and triumph are equally important.

There are some running themes, actions that will help grow more environmentally sustainable, economically viable, and socially just food and agriculture systems for us all. I call it my manifesto for food systems change: five necessary components to help us all save the world.

INVEST IN WOMEN IN AGRICULTURE

Put simply, we ignore women at our peril. Globally, women account for approximately 43 percent of the agricultural labor force, and in some countries they make up nearly 70 percent of all farmers. Yet, universally, women are not allowed access to the same resources and respect as their male counterparts.

Women face discrimination when it comes to land and livestock ownership, equal pay, participation in decision-making entities, and access to credit and financial services. Across all regions, women are less likely than men to own or control land, and the land where they grow fruits, vegetables, and other nutritious foods is often of poorer quality.

I was an emcee at the 2022 Borlaug Dialogue in Des Moines, Iowa, and the administrator for the US Agency for International Development Samantha Power was a speaker. She says, "When we hold women back, we hold everyone back."

Let me give you just one example of how this works: according to research from the UN Food and Agriculture Organization, if women farmers had the same access to resources as men, the number of hungry people in the world could be reduced by up to 150 million due to productivity gains. I've seen this on the ground with groups like the Self-Employed Women's Association (SEWA), the world's largest labor union, with more than two million members. I was able to visit SEWA farmers several years ago—about fifty women who are growing organic food and selling it under their own label to other women in urban areas. These are women who, when they have access to land, invest it back into their families. Their children go to school and receive medical care. And they've gained respect in their households and villages because they have decision-making power.

Other innovative projects are recognizing the role women play across the globe. One of my favorites is a theater group from Zimbabwe, which travels through southern Africa putting on plays that describe women's important role in communities. A women-run cooperative in Niger— made up of about fifty women farmers growing different Indigenous crops, leafy greens, ornamentals, and fruit trees—raised women farmers' incomes more than threefold.

"Women tend to bear the brunt of hunger," says Dr. Maureen Miruka, director for gender, youth, and livelihoods of CARE USA. "[Women] are also key to the solutions. . . . They are farmers, they are innovators, they are decision-makers, they are leaders, and they need to be involved."

When we invest in women, we invest not only in an individual or a group but in an entire community.

RESPECT AND HONOR INDIGENOUS PEOPLES AND PEOPLE OF COLOR IN FOOD AND AGRICULTURE

All over the world and especially in the United States, Indigenous peoples experience systemic racism, cultural appropriation, and genocide. Despite the discrimination they face, Indigenous peoples, who make up 5 percent of the global population, are protecting 80 percent of the world's remaining biodiversity. They do all this work for the planet without compensation, for the most part.

Traditional foods are the foundation of Indigenous peoples' well-being, and in many ways they are the foods of the future for all of us. These foods are resilient to pests and disease, resilient to climate change, and healthy and nutritious. And they contribute to maintaining biodiversity, something that Indigenous peoples have been doing for thousands of years in their territories.

At the 2022 COP27 Climate Change Conference in Egypt, I spent a lot of time with Indigenous leaders, like Matte Wilson of the Sičaŋǧu Food Sovereignty Initiative and Chief Caleen Sisk of the Winnemem Wintu tribe, who are thinking about how future generations can respect Indigenous practices. They are restoring traditional Indigenous foods into their communities and helping young folks understand why they are important. They believe that to go forward, we need to go back and look at why Indigenous food systems are so successful and how the world can learn from them.

In the city of Baltimore, where I live and where 65 percent of the population is Black, chefs Tonya and David Thomas are teaching eaters and young folks how to recognize and honor the Black food narrative with their work. They are recognizing the foods that those who were formerly enslaved started growing in the United States and the environmental, economic, health, and cultural benefits they still provide. That kind of remembering and honoring of people and food is more important, in my opinion, than ever before.

There need to be more spaces where the next generation of farmers, advocates, and activists learn how to care for, respect, and honor the Earth and its stewards. Like women in agriculture, they need investment. But they also need farmworker justice, land redistribution, food access with dignity, reparations, and ecological stewardship, as Leah Penniman so beautifully outlined in this book. Their land was stolen, diminishing their ability to feed themselves. They deserve more than an apology; they need actual financial compensation so that future generations can thrive.

RECOGNIZE AND SUPPORT YOUTH

Farmers all over the world are aging. The average age of a farmer in the United States is nearly sixty years old, and the same is true in parts of sub-Saharan Africa. Conversations surrounding food and farming have excluded youth voices for too long, and it's no surprise that youth all over the globe have looked at farming and our food systems as a punishment rather than an opportunity. Thankfully, that's changing.

It's not just the Greta Thunbergs of the world who are advocating for youth leadership. Groups like YPARD, an international agricultural development movement by young professionals and for young professionals, are working strategically to get young agronomists, scientists, farmers, and others at international conferences and negotiating tables so that all of us can understand what youth want and need when we're talking about the future of food.

Organizations like Slow Food are lifting young people into positions of power worldwide. In the mid-2000s, I met Edie Mukiibi in Uganda, where he was leading a school project to help students understand the importance of traditional foods—that they could be delicious and economically sustainable—and that farming is something to be respected, not looked down upon. Now, many years later, Mukiibi is the president of Slow Food International and working to improve food sovereignty and biodiversity all over the world.

In the United States, groups like 4-H and Future Farmers of America have been working for decades to help youth learn farming skills. The National Young Farmers Coalition is working to shift power and change policy to equitably resource our new generation of working farmers. Last year, they did a survey of young farmers and found that 83 percent are motivated by environmental conservation and social justice. But their survey also outlined several hurdles: 59 percent of young farmers report that it is very or extremely challenging to find affordable land. It's not surprising that 41 percent identified access to capital as a major barrier, or that 40 percent reported that managing health-care costs was very or extremely challenging.

The work of Act4Food Act4Change outlines a call to action for us all. It's a campaign that brings together youth from around the world, with the goal of providing all people with access to safe, affordable, and nutritious diets while also protecting nature, tackling climate change, and promoting human rights. These youth have developed a list of actions and are asking governments and businesses to address the broken food system: stopping and reversing land-use conversion, banning single-use plastics in food and drink packaging, and creating employment for young farmers and agri-preneurs. It's these kinds of collaborations among young folks, policymakers, and the private sector that are needed to make systemic change.

SUPPORT FOOD WORKERS AND FARMWORKERS *NOW*

It's no secret that those harvesting, slaughtering, processing, preparing, distributing, and selling food are putting their lives at risk. The COVID-19 pandemic made it clearer than ever that their essentialness can't be denied, nor can the unfair and unjust labor practices and low wages that make them vulnerable.

My hope was that as we emerged from the pandemic, we could reset. Companies that depend on food workers and farmworkers like Stonyfield, Farmer's Fridge, and Perdue Farms stepped up to implement essential benefits like hazard pay or "hero pay." Meanwhile, others regressed and suppressed workers' rights, actively thwarting their right to form unions.

Fortunately, employees at food companies across the country are saying loud and clear: business as usual is not working. As with plenty of other movements in the food system, Gen Z and young folks are leading the charge.

As of April 2023, 290 Starbucks stores in the United States have formed unions, according to Starbucks Workers United, a collective of Starbucks partners across the country. In fact, Starbucks workers organized more new unions in a twelve-month period than any company in the past twenty years. This is especially impressive given what they're up against: Starbucks workers are facing harassment and threats simply for asking for basic rights. Starbucks has fired more than two hundred organizers and permanently closed union stores, and the National Labor

Relations Board is prosecuting Starbucks with more than 1,300 separate violations of federal labor law.

One way to uplift these stories is through the power of the arts. Over the past few years, Food Tank has used theater to raise awareness of food and climate issues. Our first production, *WeCameToDance*, created by my cofounder Bernard Pollack, blended a warning of the climate crisis with a story of hope and action through food systems. The show premiered at Edinburgh Festival Fringe in 2021 and was featured in the *New York Times*. In 2023, Food Tank presented *Mermaid Coffee* (formerly *Little Peasants*), an interactive dramatic showcase of how workers at a fictional coffee chain are treated during union-organizing campaigns, at South by Southwest.

Mermaid Coffee puts audience members in baristas' shoes to demonstrate the tactics employers are using to thwart organizing efforts. The show allows us to peek behind closed doors to understand the lengths that employers like Starbucks will go to when employees ask for basic workers' rights.

"[Young people] are reviving the labor movement as a counterbalance to the power of large corporations," Richard Bensinger, an organizer with Starbucks Workers United and former national organizing director of the AFL-CIO, told Food Tank. "But for the baristas to succeed, the public and customers must hold Starbucks accountable."

As the labor movement continues to take hold in the food system, it's up to us to stand with employees and support their right to a healthy workplace free of intimidation, harassment, and threats. But as Bensinger said, it's also up to us to hold companies accountable: Starbucks and others like Amazon, Trader Joe's, Chipotle, and more who are not living up to their stated values when it comes to their treatment of workers.

We need to demand better from large companies in the food system— and that means standing with workers.

UTILIZE TRUE COST ACCOUNTING IN OUR FOOD AND AGRICULTURE SYSTEMS

The global population consumes about $9 trillion worth of food each year. But, according to a report by the UN Food Systems Summit 2021 Scientific

Group, the external cost of that food production is more than double that—nearly $20 trillion. These external costs include biodiversity loss, pollution, health-care costs and lost wages from diet-related diseases, worker abuse, poor animal welfare, and more. Unfortunately, these externalities tend to impact people of color and Indigenous peoples the most, further exacerbating the inequality and inequity that I mentioned earlier. Just one example is that Indigenous folks are nineteen times more likely to have reduced access to water and sanitation than white folks in the United States.

In addition, our food system is based on just a handful of crops like maize, soy, wheat, and rice—starchy staples that can be incredibly resource intensive to produce and that don't provide much in the way of nutrients.

The world has "created a value-destroying food system," says Roy Steiner, vice president of the food initiative at the Rockefeller Foundation. The United States creates roughly two times more economic cost than economic value from its food and agriculture systems, and similar trends can be found around the world. Steiner asks, "Who wants to be part of a value-destroying food system?" No one, right? At least, I hope not.

What if we placed value on crop and livestock systems that are healthy for people and the planet? A system that provides delicious, nutrient-dense food, that protects workers and the environment, that is regenerative and gives back more than it takes. A food system that carefully accounts for externalities and makes it more profitable to be sustainable and regenerative.

Organizations like the Sustainable Food Trust and the Rockefeller Foundation are researching how to implement true cost accounting on the ground. The idea of measuring what matters can help governments, businesses, and farmers to understand what it really costs to produce food and to make better decisions.

The Rockefeller Foundation partnered with India's Public Distribution System to supply subsidized grain to more than eight hundred million people in the country. Using true cost accounting, the foundation was able to identify hidden costs associated with greenhouse gas emissions, water use, and more. They found that the grain distribution system creates $6.1 billion per year in hidden environmental and health costs. If you can find and eliminate those externalities, you're doing more than just feeding people. You're creating a system that considers future generations and values them.

By following the advice of food policy councils to procure food for institutions like schools and hospitals locally and regionally, we could limit the transportation costs of distributing food, have more transparency in food systems, and, ultimately, provide more delicious, seasonal ingredients to students, patients, and others.

"The tools moving forward to rebuild a more localized food supply system that has equity and planetary health at its core, those opportunities are available to us now," says Paula Daniels, founding chair of the Center for Good Food Purchasing. "That's what we can do with the power [of] procurement."

GIVE RURAL FOOD SYSTEMS THE ATTENTION THEY DESERVE

Too often, farmers are blamed for these systemic issues in our food system. But, ironically, rural areas surrounded by farms experience high levels of food insecurity. According to the USDA, nearly 11 percent of people in rural areas were food insecure in 2021. A Feeding America report finds that nine of the ten counties that experience the highest rates of food insecurity are rural. In 2020, Black residents of rural counties were 2.5 times more at risk of hunger than white, non-Hispanic individuals in rural areas.

Many rural areas lack food retailers and grocery stores, making them food deserts. Broadband internet is lacking in many rural communities. A lot of people in rural America feel left out and unheard. But what if the federal government, the private sector, and researchers prioritized rural America?

As Ricardo Salvador, the director of the food and environment program for the Union of Concerned Scientists, says, "Let's invest in clean water, let's invest in clean air, and let's invest in keeping families in rural communities and building up viable rural communities." Farmers and rural communities need to be in the driver's seat and not just be recipients of agricultural research but rather partners in research. "You need to [include] the people that are going to be affected by that science and the knowledge that's being generated—the people that are actually living in rural communities," says Salvador.

In 2020, Practical Farmers of Iowa received a $1.1 million grant from the USDA's Natural Resources Conservation Service to help farmers in five states use small trains to improve fertilizer and manure management in extended crop rotations. As part of the three-year project, Practical Farmers of Iowa is working with eleven supply-chain partners including Cargill, General Mills, McDonald's, and Oatly to help farmers in Illinois, Iowa, Minnesota, Nebraska, and Wisconsin use small-grains crops and market-based solutions to lower greenhouse gas emissions linked to nitrogen from manure and fertilizer. They're helping farmers grow more diverse crops and find a market for them.

As Senator Cory Booker from New Jersey shares in this book, it won't be enough to divest power from large food businesses. We must transfer it back to the family farmers whose livelihoods have suffered the most from corporate consolidation.

Since the 1980s, Iowa counties with the most factory farm development have suffered declines across several economic indicators, including real median household income and total jobs. Farmers who are in the cycle of factory farming need help to transition, as Michael Pollan explains in Chapter 1. That is why I'm inspired by companies like Niman Ranch. The company is giving hog farmers the opportunity to raise animals that are allowed to perform their natural behaviors—and make money doing it. Niman works with 750 family farmers across the United States to raise hogs outdoors on pasture and get a premium price for their products.

Rural areas deserve far more attention and respect. This should not come at the expense of urban food security but by creating more linkages between urban and rural food systems. We need better policies that invest in rural areas. Senator Booker mentions that our current small-scale federal government programs have the potential to empower family farmers and build more resilient food systems to better serve all eaters. But this change won't happen if we don't make our voices heard on behalf of rural communities—in the Farm Bill and in every local and national election cycle.

LEVERAGE BOTH HIGH AND LOW TECH

I love my friend Christiana Musk's question in her chapter, "Does salvation lie in technology or ecology?" Of course, as Musk explores, it's not a binary choice. For Food Tank and so many of us who are looking at this issue, the answer is a blending of the two.

The COVID-19 pandemic forced us all—and especially farmers—to pivot dramatically. As farmers' markets and restaurants closed, farmers were left without outlets to sell food. Digital technology helped to change the way they do business. The Mill City Farmers Market in Minnesota set up an online store for customers to choose products from individual farms and then pick up groceries via a drive-through market. The National Young Farmers Coalition and the Glynwood Center for Regional Food and Farming held Zoom meetings to provide advice on food safety. Old-fashioned phone trees were set up so that farmers could share resources or simply check in on one another. Wholesome Wave set up an online system to deliver essentials to those who rely on Supplemental Nutrition Assistance Program (SNAP) benefits. Allowing SNAP clients to access food just as easily as eaters who order food online was a game changer, particularly for parents and the elderly.

Technology undoubtedly has a significant role to play in building a more sustainable, resilient, and equitable food system. Real-time farm data is already helping farmers use fewer inputs. Companies like Winnow are helping to significantly reduce food waste using artificial intelligence. As Michiel Bakker from Google shares, technology giants like the company he works for are making incredible strides in seafood traceability, remote monitoring and sensing, and education around climate-smart diets.

But we need a healthy amount of skepticism when new technologies come to market. We need to remember to ask some critical questions: Will an average farmer be able to adapt to the technology? What happens when farmers are not allowed the right to repair their own machinery?

Food Tank convenes the Refresh Working Group, which brings together food, agriculture, and technology experts to ensure the positive application and responsible use of emerging technologies and data across these sectors. The group represents multi-sector stakeholders from across the agriculture, food, and technology industries, including distribution, retail, and consumer goods. The group strives to ensure robust and healthy agriculture and food marketplaces where innovation thrives and where small and big players alike can drive positive improvements throughout the global food system. The group also publishes reports like the *Refresh Food + Tech Policy Platform*, which provides a broad policy framework for implementing its mission.

Of course, no technology is a silver bullet. Instead, we can combine high and low tech for real impact. Indigenous knowledge and techniques farmers have been using on the ground for centuries can be applied using modern technologies like cell phones and drones. Most of the time, we don't need to reinvent the wheel: we know what works, and our innovations can build from there.

EATING FOR CLIMATE

This book provides countless examples of the ways our food systems are making us sick and hungry and leading to multiple crises, including the climate crisis. Right now, there are zero countries that are on course to meet 2025 global nutrition targets, and global food systems contribute one-third of total human-caused greenhouse gas emissions. Something must change.

"If we continue to ignore the food systems, the 1.5 degrees Celsius goal is out of reach," says Brent Loken, global food lead scientist for the World Wildlife Fund (WWF), told me at the 2021 UN Climate Change Conference in Glasgow, Scotland. "We're not talking enough about consumption."

The global energy sector has a clearly defined road map to reach net zero by the International Energy Agency, and nearly half of the global asset-management sector has already committed to net zero by 2050 or sooner. Food and agriculture lack a clear pathway to reach net zero, and, in fact, "the food system today is net-negative in terms of the hidden costs on all of us, on climate, and on nature," says Joao Campari, global leader of food practice at WWF, "but we can't phase out food the way we are fossil fuels."

Our diets are a powerful way to build climate resilience. And we don't need to be vegan to save the world. At that same conference, UN nutrition executive secretary Stineke Oenema defined a healthy and sustainable diet as one that is primarily plant-based, including plenty of fruits, vegetables, pulses, nuts, and seeds. Notably, these diets can also contain limited amounts of red meat and a modest amount of fish and aquatic foods. Oenema also emphasized that sustainable and healthy diets are also culturally acceptable: "They are linked and rooted into the agroecological settings of a particular context, so that it's actually feasible. People need to be at the center of food system transformation, otherwise it's not going to work."

We could spend all our time and resources thinking about the health, environmental, and social impacts of our food production and consumption practices. So much onus is placed on eaters when they don't have full information or transparency.

Thankfully, there is a new generation of eaters who will force our food system to change. They are no longer willing to not see what has been hidden. They want to know the story of their food: Where was it grown, who grew it, were workers paid fairly, and what are the climate consequences of the product? These are essential questions to creating a fairer food and agriculture system, where everyone is brought to the table and no one is left behind.

FINDING YOUR PATH

When the COVID-19 pandemic revealed just how brittle our food and economic systems are, I started asking some of my personal food heroes about their vision for a food system reset. How can we mobilize for real, systemic change? What might it look like to reimagine how we grow and eat food?

For Chef Dan Barber, it's about focusing on local and regional food systems, which were built on shaking the hand of the farmer who grew your food. New models can help farmers and food producers add value to food in every possible sense—from healthy soils and nutrient-dense crops to innovative food preservation methods. A more complex value chain can mean adding more value, more nutrition, and more deliciousness to what we eat.

The private sector, specifically, has a critical role to play here: stop designing food that gives us cheap calories. Food Tank has a chief sustainability officer working group, with more than two hundred companies who are small, medium, and large. They can—and should—see a more sustainable food system as a huge opportunity, not as something that will cost them. There is a new generation of eaters that wants the story of their food, where it comes from, who grew it, and its impact on the planet, and true cost accounting gives businesses and farmers the ability to provide transparency and traceability to eaters. These leaders know that companies that can't pivot will not be around a decade from now, and I'm continuously inspired by their commitment to a more sustainable, equitable future. If you're a leader in the private industry, I urge you to get involved.

But we also need common-sense lawmaking around food and agriculture. Policymakers need to get their heads out of the sand. Food waste is just one example: if food waste were a country, it would be the third-largest emitter of greenhouse gases, after China and the United States. In the United States, the Farm Bill comes up for renewal every five years, and it's always disappointing. We need more regular conversations on Capitol Hill and in parliaments around the world about food and agriculture issues. Laws should solve the problems that actually need to be solved—the problems that farmers, eaters, and businesses face every day.

Recently, Food Tank worked with the Healthy Living Coalition to help raise awareness around the proposed Food Donation Improvement Act in the United States. Simply, it's a bill that makes it easier for individuals and institutions to donate food that would otherwise be wasted. While that seems common sense, previous legislation didn't provide oversight over who should administer the donation process or provide guidance. The Food Donation Improvement Act was an unusual piece of legislation because it had bipartisan support. Republicans and Democrats came together to solve something that is low-cost, for the most part, and can address the environmental and moral problems of food waste and help feed millions of Americans who are going hungry because of the pandemic and food price inflation. It passed on December 21, 2022. For me, this shows that the food movement does have power. It also sets the stage for more bipartisan legislation around food and agriculture—issues that should never be partisan. As Congressman Jim McGovern, whom I consider a food superhero, says: hunger should be illegal.

At the local level, many government leaders are helping to accelerate necessary changes much faster than at the national level. "With a national government that is frozen on the issue, those kinds of activities happen locally," says Steve Adler, mayor of Austin, Texas. "Urban areas are really becoming the incubators of innovation, and there's power in the city-to-city alliances."

Austin was among one hundred local governments that officially presented the Glasgow Food and Climate Declaration, a commitment by subnational governments to tackle the climate emergency through integrated food policies, during COP26. The Austin City Council also adopted the Austin Climate Equity Plan in September 2021 to equitably reach net-zero community-wide greenhouse gas emissions by 2040.

But you don't need to be working for a corporation or in government to support a more regenerative, resilient, and equitable food system for all. The nonprofit Slow Food USA hosts educational events and advocacy campaigns aiming to build solidarity through partnerships for a fairer food system. There are more than one hundred local Slow Food chapters anyone can get involved in—or start one of your own.

For those working in education, Food Tank runs an academic working group network that brings together food studies faculty to coordinate on curriculum development related specifically to sustainability. For students who want to get involved without enrolling in food systems programs, the nonprofit Food Recovery Network has chapters across forty-six states and the District of Columbia mobilizing student leaders on college campuses to work on food recovery, hunger fighting, and food justice work.

But as Daniel Katz of the Overbrook Foundation says, the best way to fight the pandemic is on Election Day. "Everything starts with one solution, and that is voting. We need an administration that is pro-environment, pro-people, and pro changing the system to rebuild a greener, safer world. We need an administration that shares that same vision."

This applies not only to those living in the United States. We can—and must—all become citizen eaters, people who vote for the kind of food system we want. And while it's important to vote with your dollar, it's also important to vote with your vote for candidates who will improve our food and agriculture systems. It's not just at the national level but at the level of local school boards, credit unions, and mayoral races—or run for office yourself. I've been meeting people in their twenties who are farmers or food advocates and are becoming local politicians because they want food procurement to change, or they want more focus on solving the climate crisis. They're the next generation of leaders.

What is clear to me is that we don't have time to waste. We're facing multiple crises: the climate crisis, the biodiversity loss crisis, the public health crisis. We—and I mean all of humanity—have been cultivating our own food for about ten thousand years. For most of that time, we've been spoiled. There weren't that many of us, and there was plenty to live from. That abundance has tended to make us lazy. It made us think that the Earth is expendable. It's not. And that illusion and laziness can't last.

Now there are simply too many of us. To put it into context, if you were to total up the people that have lived in all the ten thousand years since we domesticated plants, more than one in fourteen of us woke up this morning. Seven percent of everyone who has ever depended on a farmer for food is alive right now. That's a huge number. Population scientists say we'll top off at ten billion people on this planet in about thirty years—and last year we passed eight billion. The time when we could take environmental, economic, and social sustainability—or rural areas—for granted is over. That's the bad news.

The good news is that we still have time. There is time to realize that what we've taken for granted is not guaranteed. We can get back on track. Humanity is still young. I said we're 7 percent of everyone that's lived since farming began, but if humans survive another five thousand years, all our farming ancestors and all of us combined will account for just 10 percent of human history. It boggles my mind every time I think about these numbers.

As Oxford professor of philosophy William MacAskill wrote in a recent article, "We are the ancients." Unlike anyone before us, and just like everyone who will come after, we must discover how to live on a full planet. We need to start thinking and behaving like the future's ancestors, or we won't be.

17.

To Learn More and to Get Involved

Are you interested in learning more about the food system and the ways it can be reformed? Are you looking for ideas about how you can improve the way your own family eats? Are you wondering how you can get involved as an activist or advocate in support of creating a healthier, fairer, and more sustainable food system in your community, nation, and world? Whatever your learning objective may be, here is a collection of resources—advocacy groups, websites, educational organizations, and more—that you will find helpful.

Our list is divided by topic for easy reference, but you probably should glance at the entire thing even if you have a specific interest in mind, since the connections among issues in the food world are enormous and vital. Also note that our list emphasizes sources of national and global scope, but there are many local, state, and regional organizations engaged in complementary work; consider seeking out one or more of these groups for opportunities to get involved in your own community.

SUSTAINABLE AGRICULTURE

The Story So Far
Organic foods have become more accessible than ever, but the integrity of the movement has come into question. Regenerative agriculture has risen as the next big thing in organics, and the connection between conventional foodways and the climate crisis is becoming more widely understood.

To Learn More and to Get Involved

American Farmland Trust is a nonprofit organization dedicated to sustainable agricultural practices, protecting farmlands, and supporting small farmers. https://farmland.org/

Black Oaks Center is an eco-campus, farm, and school in rural Illinois dedicated to revitalizing sustainable farming in a historic Black farming community. www.blackoakscenter.org/

Demeter Association, Inc., seeks to heal the planet through agriculture by enabling people to farm successfully, in accordance with biodynamic practices and principles. www.demeter-usa.org/about-demeter/

Food & Water Watch is a nonprofit organization promoting policies to make agriculture more sustainable and alleviate the damage caused by climate change. www.foodandwaterwatch.org

Land Core is a 501(c)(3) organization with a mission to advance soil health policies and programs that create value for farmers, businesses, and communities. https://landcore.org

The National Sustainable Agriculture Coalition is an alliance of grassroots organizations that advocates for federal policy reform to advance the sustainability of agriculture, food systems, natural resources, and rural communities. https://sustainableagriculture.net/about-us/

The Regenerative Organic Alliance aims to help build a healthy food system that respects land and animals, empowers people, and restores communities and ecosystems through its certification program for regenerative organic farming. https://regenorganic.org

The Rodale Institute's California Organic Center serves California farmers interested in regenerative agriculture by solving challenges, conducting regionally significant research, and serving as a hub for education and extension. https://rodaleinstitute.org

FOOD AND AGRICULTURE TECHNOLOGY

The Story So Far

Investments in good food solutions have exploded in recent years. Some agriculture innovators are turning to new technologies for alternative methods of growing food, including vertical farming, aquaponics, and hydroponics, as well as better ways of reducing food waste.

To Learn More and to Get Involved

EAT is a global nonprofit start-up dedicated to transforming our food system through sound science, impatient disruption, and novel partnerships with foundations, academic institutions, organizations, and companies. https://eatforum.org/about/

Food System 6 is a business accelerator that focuses on supporting food entrepreneurs who support economic and racial justice, environmental sustainability, rebuilding local food systems, and community health. www.foodsystem6.org

Food+Tech Connect provides a website, newsletters, and consulting services that help people understand the top food-tech, investment, and innovation trends, thereby promoting the transformation of the food system. https://foodtechconnect.com/about/

The Good Food Institute is a nonprofit think tank and international network of organizations working to accelerate alternative protein innovation through technologies like plant-based meat and meat grown from cells. https://gfi.org

New Harvest is a donor-funded, nonprofit research institute advancing the science behind cultured meat. https://new-harvest.org

OpenTEAM is a collaborative community of farmers, ranchers, scientists, engineers, and food companies creating tools for improving soil health and helping agriculture address the problems of climate change. https://openteam.community

BIG FOOD AND BIG AG

The Story So Far
Corporations in the food and agricultural industries are consolidating at an alarming rate, with mergers and acquisitions narrowing the majority stake in our food system. Governmental deregulation has made it harder for small companies to survive and compete, reducing the diversity and sustainability of the food system.

To Learn More and to Get Involved
Agrarian Trust is a nonprofit organization working to create agricultural commons that support small farmers, including people from disempowered groups, in practicing sustainable farming methods that increase equality and justice. www.agrariantrust.org/about/

Black Family Land Trust, Inc., is a North Carolina–based conservation land trust dedicated to the preservation and protection of African American and other historically underserved landowners' assets. www.bflt.org

Black Urban Growers is a nonprofit organization that sponsors an annual conference dedicated to the empowerment and resilience of Black agriculture worldwide, with the specific goal of creating more equitable and sustainable food systems. www.blackurbangrowers.org

The California Farmer Justice Collaborative works to build a fairer food system for small farmers, especially those from underrepresented communities. www.farmerjustice.com

The Campaign for Family Farms and the Environment is an advocacy group representing four Midwest state-based membership organizations and two national organizations fighting against corporate factory farms. www.iatp.org/campaign-family-farms-and-environment

The Federation of Southern Cooperatives advocates for Black farmers and other small landowners through cooperative economic development. www.federation.coop

ANIMAL WELFARE

The Story So Far

Animal welfare activists are pursuing a two-pronged approach to minimizing suffering: alternatives to killing animals for meat (including cell-cultured meat and the continued proliferation of meat alternatives) and improving the current standards of both life and death for farm animals.

To Learn More and to Get Involved

The Animal Agriculture Reform Collective is a nonprofit organization working to create an animal agriculture system that is sustainable, is fair to workers, protects communities, and practices humane treatment of animals. https://multiplier.org/project/animal-agriculture-reform-collaborative/#

Farm Forward is an advocacy group whose goal is to end factory farming by educating the public and policymakers about the benefits of plant-based foods and pushing for high standards of animal welfare in agriculture. www.farmforward.com

Food Animal Concerns Trust works to protect animal welfare through policy advocacy and training programs for independent farmers. www.foodanimalconcernstrust.org

Mercy for Animals is a nonprofit organization working to end factory farming by advocating for nonanimal protein sources and promoting policy change at all levels. https://mercyforanimals.org

HEALTHY EATING

The Story So Far

Despite the United States' new focus on healthy eating, obesity has continued to increase and has skyrocketed in other countries. Still, consumer preferences have changed, with an emphasis on "conscious consumption." The popularity and availability of plant-based diets and foods have risen exponentially, and new food companies aimed at satisfying evolving consumer demands are popping up.

To Learn More and to Get Involved

The Edible Schoolyard Project is a nonprofit organization that works to provide free, sustainable, healthy meals to schoolkids while supporting farmers who take care of the land and their workers. https://edibleschoolyard.org

The Food for Climate League works to connect the dots between sustainable food culture and people's unique needs, values, and cultures through research, storytelling, and education. www.foodforclimateleague.org

The Lexicon is a nonprofit organization that uses storytelling in many forms to identify and accelerate the adoption of practices that will build more resilient agri-food systems and help combat climate change. www.thelexicon.org

The Museum of Food and Drink is a New York–based "museum on wheels" dedicated to educating the public about the intimate connections among human cultures and the foods we eat. www.mofad.org

ADVOCACY, POWER, AND POLITICS

The Story So Far

New waves of food advocacy have led to activists' presence at every stage of the food supply chain: how it's produced, who gets to produce it, how it's sold, and who gets to eat it. Farmworkers, food producers, service employees, family farmers, underrepresented demographic groups, consumer advocates, and environmental activists are all demanding roles in shaping a better food system.

To Learn More and to Get Involved

The Climate, Animal, Food, and Environmental Law and Policy Lab (CAFE Lab) at Yale Law School seeks to develop innovative law and policy initiatives to reform the practices of industrial food producers around worker rights, the environment, food health and safety, and more. https://law.yale.edu/animals/initiatives/cafe-law-policy -lab

The Coalition of Immokalee Workers is a worker-based human rights organization fighting for fair wages, an end to human trafficking, and protection of other basic rights for farmworkers. https://ciw-online .org/about/

Farm Action is a research, policy, and advocacy organization that seeks to protect workers, the environment, and rural communities from the damage caused by industrial agriculture. https://farmaction.us /about-us/

A Growing Culture is a nonprofit organization dedicated to supporting the global food sovereignty movement. www.agrowingculture .org/#

The Intertribal Agriculture Council conducts a wide range of programs designed to further the goal of improving Indian agriculture to maximize resources for tribal members. www.indianag.org

The National Black Food and Justice Alliance is a coalition of Black-led organizations developing Black leadership, supporting Black communities, organizing Black self-determination, and building institutions for Black food sovereignty and liberation. www.blackfoodjustice .org

The National Young Farmers Coalition is an organization of farmers and ranchers who steward the struggle to transform agriculture, working for justice and collective liberation of our food and farm systems. www.youngfarmers.org/about/

The Public Justice Food Project uses litigation, base building, and storytelling to address the structural and institutional inequities in the current food system and to build a system that is free from exploitation and extraction. https://food.publicjustice.net

Rural Coalition works to support farmworkers and farm communities from diverse backgrounds through a program of public policy monitoring, technical assistance and capacity building, participatory collaborative research, and education. www.ruralco.org/

NOTES

CHAPTER 1: THE SICKNESS IN OUR FOOD SUPPLY

1. This history is recounted in Barry C. Lynn, *Cornered: The New Monopoly Capitalism and the Economics of Destruction* (Hoboken, NJ: Wiley, 2011), 135–138.

2. Claire Kelloway, "Why Are Farmers Destroying Food While Grocery Stores Are Empty?," *Washington Monthly*, April 28, 2020.

3. "In America, the Virus Threatens a Meat Industry That Is Too Concentrated," *Economist*, April 30, 2020.

4. Leah Douglas, "Mapping COVID-19 in the Food System," Food and Environment Reporting Network (FERN), April 22, 2020. FERN has covered this story extensively and compiled statistics. Also see Esther Honig and Ted Genoways, " 'The Workers Are Being Sacrificed': As Cases Mounted, Meatpacker JBS Kept People on Crowded Factory Floors," *Mother Jones*, May 1, 2020. *Civil Eats*, FERN, and *Mother Jones* have done an excellent job of covering the outbreaks in the meat industry.

5. Magaly Licolli, "As Tyson Claims the Food Supply Is Breaking, Its Workers Continue to Suffer," *Civil Eats*, April 30, 2020.

6. Tyler Whitley, "Don't Blame Farmers Who Have to Euthanize Their Animals. Blame the Companies They Work For," *Civil Eats*, April 30, 2020.

7. It's worth remembering that the federal government actively promotes meat consumption in myriad ways, from USDA advertising campaigns—"Beef: It's What's for Dinner"—to exempting feedlots from provisions of the Clean Water and Clean Air Acts, to the dietary guidelines it issues and the heavy subsidies it gives for animal feed.

8. See, for example, Daniel A. Medina, "As Amazon, Walmart, and Others Profit Amid Coronavirus Crisis, Their Essential Workers Plan Unprecedented Strike," *Intercept*, April 28, 2020.

9. Shikha Garg et al., "Hospitalization Rates and Characteristics of Patients Hospitalized with Laboratory-Confirmed Coronavirus Disease 2019, COVID-NET, 14 States, March 1–30, 2020," *Morbidity and Mortality Weekly Report* 69, no. 15 (April 17, 2020).

10. Xiaomin Luo et al., "Prognostic Value of C-Reactive Protein in Patients with COVID-19," medRxiv, March 23, 2020. The study has not yet been peer-reviewed.

CHAPTER 3: FOOD, COOKING, MEALS, GOOD HEALTH, AND WELL-BEING

1. "2025 Advisory Committee," Dietary Guidelines for Americans, USDA and HHS, www.dietaryguidelines.gov/2025-advisory-committee.

2. "History of the *Dietary Guidelines*," Dietary Guidelines for Americans, USDA and HHS, www.dietaryguidelines.gov/about-dietary-guidelines/history-dietary-guidelines.

3. Carlos A. Monteiro, "Nutrition and Health: The Issue Is Not Food, nor Nutrients, So Much as Processing," *Public Health Nutrition* 12, no. 5 (May 2009): 729–731, https://doi.org/10.1017/S1368980009005291.

4. Monteiro, "Nutrition and Health."

5. Renata Bertazzi Levy-Costa et al., "Household Food Availability in Brazil: Distribution and Trends, 1974–2003," *Revista de Saúde Pública* 39, no. 4 (2005): 1–10, www.fsp.usp.br/rsp.

6. Michael Pollan, *In Defense of Food: An Eater's Manifesto* (New York: Penguin Press, 2008).

7. Geoffrey Cannon, "Out of the Box," *Public Health Nutrition* 12, no. 5 (May 2009): 732–733, https://doi.org/10.1017/S1368980009005370.

8. Ministry of Health of Brazil, *Dietary Guidelines for the Brazilian Population*, 2014, English edition, 2016.

9. Carlos Augusto Monteiro et al., *Ultra-processed Foods, Diet Quality, and Health Using the NOVA Classification System* (Rome: UNFAO, 2019), www.fao.org/3/ca5644en/ca5644en.pdf.

10. UNFAO and International Fund for Agricultural Development (IFAD), *United Nations Decade of Family Farming 2019–2028: Global Action Plan* (Rome: UNFAO, 2019), www.fao.org/3/ca4672en/ca4672en.pdf; Simon Blondeau and Anna Korzenszky, *Family Farming* (Rome: UNFAO, 2022), www.fao.org/documents/card/en/c/cb8227en/.

11. Corinna Hawkes, "The Role of Foreign Direct Investment in the Nutrition Transition," *Public Health Nutrition* 8, no. 4 (2005): 357–365, https://doi.org/10.1079/PHN2004706; Organisation for Economic Co-operation and Development (OECD), *Foreign Direct Investment for Development: Maximizing Benefits, Minimizing Costs*, 2002, www.oecd.org/investment/investmentfordevelopment/1959815.pdf; OECD, "FDI in Figures," April 2016, www.oecd.org/corporate/FDI-in-Figures-April-2016.pdf.

12. Carlos A. Monteiro and Geoffrey Cannon, "The Impact of Transnational 'Big Food' Companies on the South: A View From Brazil," *PLoS Medicine* 9, no. 7 (July 3, 2012), https://doi.org/10.1371/journal.pmed.1001252.

13. Daniela Martini et al., "Ultra-processed Foods and Nutritional Dietary Profile: A Meta-analysis of Nationally Representative Samples," *Nutrients* 13, no. 10 (2021): 3390, https://doi.org/10.3390/nu13103390.

14. Daniela Neri et al., "Ultraprocessed Food Consumption and Dietary Nutrient Profiles Associated with Obesity: A Multicountry Study of Children and Adolescents," *Obesity Reviews* 23, no. S1 (January 2022), https://doi.org/10.1111/obr.13387.

15. Junxiu Liu et al., "Consumption of Ultraprocessed Foods and Diet Quality Among US Children and Adults," *American Journal of Preventive Medicine* 62, no. 2 (February 2022): 252–264, https://doi.org/10.1016/j.amepre.2021.08.014.

16. Samuel J. Dicken and Rachel L. Batterham, "The Role of Diet Quality in Mediating the Association Between Ultra-processed Food Intake, Obesity and Health-Related Outcomes: A Review of Prospective Cohort Studies," *Nutrients* 14, no. 1 (2022): 23, https://doi.org/10.3390/nu14010023.

17. Dicken and Batterham, "The Role of Diet Quality."

18. Lihui Zhou et al., "Impact of Ultra-processed Food Intake on the Risk of COVID-19: A Prospective Cohort Study," *European Journal of Nutrition* 62, no. 1 (2023): 275–287, https://doi.org/10.1007/s00394-022-02982-0.

19. Zhou et al., "Impact of Ultra-processed Food"; Felipe Mendes Delpino et al., "Ultra-processed Food and Risk of Type 2 Diabetes: A Systematic Review and Meta-analysis of Longitudinal Studies," *International Journal of Epidemiology* 51, no. 4 (August 2022): 1120–1141, https://doi.org/10.1093/ije/dyab247; Wanich Suksatan et al., "Ultra-processed Food Consumption and Adult Mortality Risk: A Systematic Review and Dose-Response Meta-analysis of 207,291 Participants," *Nutrients* 14, no. 1 (2022): 174, https://doi.org/10.3390/nu14010174.

20. Kevin D. Hall et al., "Ultra-processed Diets Cause Excess Calorie Intake and Weight Gain: An Inpatient Randomized Controlled Trial of Ad Libitum Food Intake," *Cell Metabolism* 30, no. 1 (2019): 67–77, https://doi.org/10.1016/j.cmet.2019.05.008.

21. Gyorgy Scrinis and Carlos Augusto Monteiro, "Ultra-processed Foods and the Limits of Product Reformulation," *Public Health Nutrition* 21, no. 1 (2018): 247–252, https://doi.org/10.1017/S1368980017001392.

22. Claus Leitzmann, "Characteristics and Health Benefits of Phytochemicals," *Forschende Komplementärmedizin* 23, no. 2 (2016): 69–74, https://doi.org/10.1159/000444063; Christine Morand and Francisco A. Tomás-Barberán, "Contribution of Plant Food Bioactives in Promoting Health Effects of Plant Foods: Why Look at Interindividual Variability?," *European Journal of Nutrition* 58 (2019): 13–19, https://doi.org/10.1007/s00394-019-02096-0.

23. Ciarán G. Forde, Monica Mars, and Kees de Graaf, "Ultra-processing or Oral Processing? A Role for Energy Density and Eating Rate in Moderating Energy Intake from Processed Foods," *Current Developments in Nutrition* 4, no. 3 (March 2020): nzaa019, https://doi.org/10.1093/cdn/nzaa019.

24. Forde, Mars, and de Graaf, "Ultra-processing or Oral Processing?"; Filippa Juul, Georgeta Vaidean, and Niyati Parekh, "Ultra-processed Foods and Cardiovascular

Diseases: Potential Mechanisms of Action," *Advances in Nutrition* 12, no. 5 (September 2021): 1673–1680, https://doi.org/10.1093/advances/nmab049.

25. Linda Geddes, "Go with Your Gut: Scientist Tim Spector on Why Food Is Not Just Fuel," *Guardian*, May 15, 2022, www.theguardian.com/lifeandstyle/2022/may/15 /go-with-your-gut-tim-spector-power-of-microbiome; Brigitte M. González Olmo, Michael J. Butler, Ruth M. Barrientos, "Evolution of the Human Diet and Its Impact on Gut Microbiota, Immune Responses, and Brain Health," *Nutrients* 13, no. 1 (2021): 196, https://doi .org/10.3390/nu13010196.

26. "Food Additives," WHO, January 31, 2018, www.who.int/news-room/fact-sheets /detail/food-additives; Eloi Chazelas et al., "Food Additives: Distribution and Co-occurrence in 126,000 Food Products of the French Market," *Scientific Reports* 10 (2020): 3980, https:// doi.org/10.1038/s41598-020-60948-w.

27. USDA and HHS, *Dietary Guidelines for Americans: 2020–2025* (Washington, DC: USDA, 2020), www.dietaryguidelines.gov/sites/default/files/2021-03/Dietary_Guidelines _for_Americans-2020-2025.pdf.

28. Eloi Chazelas et al., "Exposure to Food Additive Mixtures in 106,000 French Adults from the NutriNet-Santé Cohort," *Scientific Reports* 11 (2021), https://doi.org/10.1038 /s41598-021-98496-6.

29. Eric Schlosser, "Why the Fries Taste Good," chap. 5 in *Fast Food Nation: What the All-American Meal Is Doing to the World* (Boston, MA: Houghton Mifflin, 2001).

30. Juul, Vaidean, and Parekh, "Ultra-processed Foods and Cardiovascular Diseases."

31. Juul, Vaidean, and Parekh, "Ultra-processed Foods and Cardiovascular Diseases"; Lisa Lefferts, *Obesogens: Assessing the Evidence Linking Chemicals to Food to Obesity* (Washington, DC: Center for Science in the Public Interest, 2023), www.cspinet.org/sites/default /files/2023-02/CSPI_Obesogens_Report_2-2023.pdf.

32. Michael Moss, *Hooked: Food, Free Will, and How the Food Giants Exploit Our Addictions* (New York: Random House, 2021); Ashley N. Gearhardt and Erica M. Schulte, "Is Food Addictive? A Review of the Science," *Annual Review of Nutrition* 41 (2021): 387–410, https://doi.org/10.1146/annurev-nutr-110420-111710.

33. Gearhardt and Schulte, "Is Food Addictive?"

34. Theodosius Dobzhansky, "Nothing in Biology Makes Sense Except in the Light of Evolution," *American Biology Teacher* 35, no. 3 (1973): 125–129, https://doi .org/10.2307/4444260; Richard W. Wrangham, "The Evolution of Human Nutrition," *Current Biology* 23 (2013): R354–R355, https://doi.org/10.1016/j.cub.2013.03.061; Ann Gibbons, "The Evolution of Diet," *National Geographic*, 2014, www.nationalgeographic.com /foodfeatures/evolution-of-diet/.

35. Monteiro et al., *Ultra-processed Foods, Diet Quality, and Health*; Claude Fischler, "The 'McDonaldization' of Culture," in *Food: A Culinary History*, eds. Jean-Louis Flandrin and Massimo Montanari, trans. Albert Sonnenfeld (New York: Columbia University Press, 2013).

36. Jean-Jacques Boutaud, Anda Becuț, and Angelica Marinescu, "Food and Culture: Cultural Patterns and Practices Related to Food in Everyday Life," *International Review of Social Research* 6, no. 1 (2016): 1–3, https://doi.org/10.1515/irsr-2016-0001.

37. UNFAO and IFAD, *United Nations Decade of Family Farming*; Blondeau and Korzenszky, *Family Farming*; UNFAO, "FAO Unveils New Technical Platforms for Family Farming," news release, December 2, 2021, https://reliefweb.int/report/world /fao-unveils-new-technical-platform-family-farming.

38. Anthony Fardet and Edmond Rock, "Ultra-processed Foods and Food System Sustainability: What Are the Links?," *Sustainability* 12, no. 15 (2020): 6280, https://doi.org /10.3390/su12156280.

39. Monteiro and Cannon, "The Impact of Transnational 'Big Food.'"

40. Klaus Michael Meyer-Abich, "Human Health in Nature—Towards a Holistic Philosophy of Nutrition," *Public Health Nutrition* 8, no. 6A (2005): 738–782, https://doi .org/10.1079/PHN2005788; Geoffrey Cannon, "The Rise and Fall of Dietetics and of Nutrition Science, 4000 BCE–2000 CE," *Public Health Nutrition* 8, no. 6A (2005): 701–705, https://doi.org/10.1079/PHN2005766.

41. Ministry of Health of Brazil, *Dietary Guidelines*.

42. Chris Arsenault, "Free School Meals in Brazil Help Local Farmers Stay on the Land," Reuters, August 30, 2016, www.reuters.com/article/brazil-landrights-politics-idUSL8N1 B433N.

43. "Regions," Food-Based Dietary Guidelines, UNFAO, www.fao.org/nutrition/education /food-dietary-guidelines/regions/en/; Israeli Ministry of Health, *Nutritional Recommendations*, 2019, www.health.gov.il/PublicationsFiles/dietary%20guidelines%20EN.pdf; National Coordinating Committee on Food and Nutrition, *Malaysian Dietary Guidelines 2020* (Putrajaya: Ministry of Health Malaysia, 2021), https://hq.moh.gov.my/nutrition/wp -content/uploads/2021/07/Web%20MDG.pdf.

44. Le Haut Conseil de la Santé Publique, "Relatif Aux Objectifs de Santé Publique Quantifiés Pour La Politique Nutritionnelle de Santé Publique (PNNS) 2018–2022" [Quantified public health objectives for public health nutrition policy (PNNS) 2018–2022], February 9, 2018, www.hcsp.fr/explore.cgi/avisrapportsdomaine?clefr=648.

45. Alice H. Lichtenstein et al., "2021 Dietary Guidance to Improve Cardiovascular Health: A Scientific Statement from the American Heart Association," *Circulation* 144, no. 23 (December 7, 2021): e472–e487, https://doi.org/10.1161/CIR.0000000000001031.

46. Tom H. Karlsen et al., "The EASL–*Lancet* Liver Commission: Protecting the Next Generation of Europeans Against Liver Disease Complications and Premature Mortality," *Lancet* 399 (2021): 61–116, https://doi.org/10.1016/S0140-6736(21)01701-3.

47. Ministry of Health of Brazil, *Dietary Guidelines for Brazilian Children Under 2 Years of Age*, 2021.

48. José María Bengoa et al., "Nutritional Goals for Health in Latin America," *Food and Nutrition Bulletin* 11, no. 1 (1989): 4–20, https://archive.unu.edu/unupress/food/8F111e/8F111E01.htm.

49. Ministério da Saúde, *Guia Alimentar Para a População Brasileira*, 2005, 2008.

50. Susanne Kerner, Cynthia Chou, and Morten Warmind, eds., *Commensality: From Everyday Food to Feast* (London: Bloomsbury, 2015).

51. Claude Lévi-Strauss, "The Culinary Triangle," in *Food and Culture: A Reader*, eds. Carole Counihan and Penny Van Esterik (London: Routledge, 1997).

52. Dan Buettner, *The Blue Zones: Lessons for Living Longer from the People Who've Lived the Longest* (Washington, DC: National Geographic Books, 2008).

53. "Mediterranean Diet," Representative List of the Intangible Cultural Heritage of Humanity, UNESCO, 2013, https://ich.unesco.org/en/RL/mediterranean-diet-00884.

54. Claudia Roden, *A Book of Middle Eastern Food* (London: Penguin, 1968).

55. Emma Rothschild, "Hangchow Retrouvé," *London Review of Books*, May 22, 1980, www.lrb.co.uk/the-paper/v02/n10/emma-rothschild/hangchow-retrouve.

56. Mark L. Wahlqvist, Gayle Savige, and Naiyana Wattanapenpaiboon, "Cuisine and Health: A New Initiative for Science and Technology: 'The Zhejiang Report' from Hangzhou," *Asia Pacific Journal of Clinical Nutrition* 13, no. 2 (2004): 121–124.

57. Rob Moodie et al., "Profits and Pandemics: Prevention of Harmful Effects of Tobacco, Alcohol, and Ultra-processed Food and Drink Industries," *Lancet* 381, no. 9867 (February 23, 2013): 670–679, https://doi.org/10.1016/S0140-6736(12)62089-3; Pan American Health Organization (PAHO), *Ultra-processed Food and Drink Products in Latin America: Trends, Impact on Obesity, Policy Implications* (Washington, DC: PAHO and WHO, 2015), https://iris.paho.org/bitstream/handle/10665.2/7699/9789275118641_eng.pdf; Marco Springmann et al., "Analysis and Valuation of the Health and Climate Change Co-benefits of Dietary Change," *PNAS* 113, no. 15 (2016): 4146–4151, https://doi.org/10.1073/pnas.1523119113; "Fiscal and Pricing Policies: Evidence Report and Framework," Public Health England, December 11, 2018, www.gov.uk/government/publications/fiscal-and-pricing-policies-evidence-report-and-framework; Carlos Augusto Monteiro et al., "The Need to Reshape Global Food Processing: A Call to the United Nations Food Systems Summit," *BMJ Global Health* 6 (2021), https://doi.org/10.1136/bmjgh-2021-006885; Barry M. Popkin et al., "Towards Unified and Impactful Policies to Reduce Ultra-processed Food Consumption and Promote Healthier Eating," *Lancet: Diabetes and Endocrinology* 9 (2021): 462–470, https://doi.org/10.1016/S2213-8587(21)00078-4.

CHAPTER 4: DAIRY TOGETHER

1. "Wisconsin Monthly Dairy Farms Statistics," Farm and Dairy Statistics, Proudly Cheese Wisconsin, accessed February 5, 2023, www.wisconsincheese.com/media/facts-stats/farm-dairy-statistics.

2. "Milk Cows" and "Milk Production," National Agricultural Statistics Services, USDA, accessed April 12, 2023, www.nass.usda.gov/Charts_and_Maps/Milk_Production _and_Milk_Cows/.

3. For a primer on this issue: Mary Hendrickson, Philip H. Howard, and Douglas Constance, "Power, Food and Agriculture: Implications for Farmers, Consumers and Communities" (working paper, University of Missouri, College of Agriculture, Food, and Natural Resources, Division of Applied Social Sciences, November 1, 2017), https://dx.doi.org/10.2139/ssrn.3066005.

4. Russell Redman, "Are We on the Verge of a New Grocery Landscape?," *Supermarket News*, December 6, 2022, www.supermarketnews.com/retail-financial/are-we-verge-new -grocery-landscape; Shelby Vittek, "Why Supermarket Monopolies Are Bad for the Farm Economy," Ambrook Research, January 20, 2023, https://ambrook.com/research/kroger -albertsons-merger-farmer-opposition.

5. Isabella Simonetti and Julie Creswell, "Food Prices Soar, and So Do Companies' Profits," *New York Times*, November 1, 2022, www.nytimes.com/2022/11/01/business/food-prices -profits.html.

6. "U.S. Dairy Exports—Percent of Production, 1996–2020," US Dairy Export Council, accessed April 12, 2023, www.usdec.org/assets/Images/Supplier%20Site%20Images/Research andMarkets/GlobalDairyMarketOutlook/TopCharts/percent%20production(0).png.

7. I look to the wise people at the Institute for Agriculture and Trade Policy (IATP) for good analysis on how we can build just and sustainable trade policies that do not pit farmers in one country against those in another. See Kristin Dawkins, "Balancing Policies for Just and Sustainable Trade," IATP, January 15, 1994, www.iatp.org/documents/balancing-policies -for-just-and-sustainable-trade-1.

8. Aldo Leopold, *A Sand County Almanac* (New York: Oxford University Press, 1968).

9. "Annual Milk Production Overview for 2021," Proudly Cheese Wisconsin, accessed February 12, 2023, www.wisconsincheese.com/media/facts-stats/farm-dairy-statistics.

10. "Dairy Together Meetings Spark Cautious Optimism," Dairy Together, accessed April 12, 2023, www.dairytogether.com/single-post/2018/03/21/Dairy-Together-meetings-spark -cautious-optimism.

11. Bob Gray, "The Dairy Summit: The Meeting in Albany Attracts Some 400 from Across the Industry and Across the Country," *Dairy Business*, August 17, 2018, www.dairybusiness .com/the-dairy-summit-the-meeting-in-albany-attracts-some-400-from-across-the-industry -and-across-the-country/.

12. Chuck Nicholson and Mark Stephenson, "Analyses of Proposed Alternative Growth Management Programs for the US Dairy Industry," Dairy Together, video, accessed April 12, 2023, www.dairytogether.com/videos.

13. "The Dairy Revitalization Plan: A Vision for the Future of Dairy," Dairy Together, accessed April 12, 2023, www.dairytogether.com/_files/ugd/629d75_9d6828bb745e4ceb83 a18e1ef4b1da2a.pdf.

CHAPTER 5: FOOD AND LAND JUSTICE

1. "Food Deserts in the United States," Annie E. Casey Foundation, February 13, 2021, www.aecf.org/blog/exploring-americas-food-deserts.

2. USDA, "United States Farms with American Indian or Alaska Native Producers," 2017 Census of Agriculture, accessed April 12, 2023, www.nass.usda.gov/Publications /AgCensus/2017/Online_Resources/Race,_Ethnicity_and_Gender_Profiles/cpd99000.pdf.

3. USDA, "United States Farms."

4. Amanda Gold et al., *Findings from the National Agricultural Workers Survey, 2019–2020* (North Bethesda, MD: JBS International, 2022), www.dol.gov/sites/dolgov/files /ETA/naws/pdfs/NAWS Research Report 16.pdf; "Labor Force Statistics from the Current Population Survey," US Bureau of Labor Statistics, accessed April 12, 2023, www.bls.gov/cps /cpsaat11.htm.

5. United Farm Workers and Bon Appétit Management Company Foundation, *Inventory of Farmworker Issues and Protections in the United States*, 2011, https://s3.amazonaws.com /oxfam-us/static/oa3/files/inventory-of-farmworker-issues-and-protections-in-the-usa.pdf.

6. "Farmworker Health Factsheet: Demographics," National Center for Farmworker Health, September 2012, www.ncfh.org/uploads/3/8/6/8/38685499/fs-migrant_demographics.pdf.

7. "Low Wages," Issues Affecting Farm Workers, National Farm Worker Ministry, last modified September 2022, http://nfwm.org/education-center/farm-worker-issues/low-wages/.

8. Hossein Ayazi and Elsadig Elsheikh, *The US Farm Bill: Corporate Power and Structural Racialization in the United States Food System* (Berkeley, CA: Hass Institute, 2015), http://haasinstitute.berkeley.edu/sites/default/files/haasinstitutefarmbillreport_publish_0.pdf.

9. United Farm Workers and Bon Appétit Management Company Foundation, *Inventory*.

10. *Hungry for Justice* (blog), Food Justice Certification, Agricultural Justice Project, accessed April 12, 2023, www.agriculturaljusticeproject.org/en/blog/.

11. "Our Work," Agriculture and Land-Based Training Association, accessed April 12, 2023, https://albafarmers.org/our-work/; Groundswell Center for Local Food & Farming, accessed April 12, 2023, https://groundswellcenter.org/; Soul Fire Farm, accessed April 12, 2023, www.soulfirefarm.org.

12. Mia de Graaf, "How the West Was Stolen: Scale of Native American Dispossession Revealed in Striking Time-Lapse Video," *Daily Mail*, January 8, 2015, www.dailymail.co .uk/news/article-2902380/Story-Native-American-dispossession-told-unforgettable-new -visualizations.html.

13. Anna Deen, "What Is Heirs' Property? A Huge Contributor to Black Land Loss You Might Not Have Heard Of," Fix, March 17, 2021, https://grist.org/fix/justice/what -is-heirs-property-a-huge-contributor-to-black-land-loss-you-might-not-have-heard-of/.

14. USDA, "2017 Race, Ethnicity and Gender Profiles," 2017 Census of Agriculture, accessed April 12, 2023, www.nass.usda.gov/Publications/AgCensus/2017/Online_Resources /Race,_Ethnicity_and_Gender_Profiles/.

15. Pamela Browning et al., *The Decline of Black Farming in America* (Washington, DC: Commission on Civil Rights, 1982), https://eric.ed.gov/?id=ED222604.

16. Richard Fry, Jesse Bennett, and Amanda Barroso, "Racial and Ethnic Gaps in the U.S. Persist on Key Demographic Indicators," Pew Research Center, January 12, 2021, www.pewresearch.org/interactives/racial-and-ethnic-gaps-in-the-u-s-persist-on-key-demographic-indicators/.

17. Chris McGreal, "US Should Return Stolen Land to Indian Tribes, Says United Nations," *Guardian*, May 4, 2012, www.theguardian.com/world/2012/may/04/us-stolen-land-indian-tribes-un.

18. "Lisjan," Sogorea Te' Land Trust, accessed April 12, 2023, https://sogoreate-landtrust.org/lisjan/.

19. Farm Service Agency, "Native American Tribal Loans," USDA, accessed April 12, 2023, www.fsa.usda.gov/programs-and-services/farm-loan-programs/native-american-loans/index.

20. "Our Vision," Northeast Farmers of Color Land Trust, accessed April 12, 2023, https://nefoclandtrust.org/our-work.

21. Black Family Land Trust, Inc., accessed April 12, 2023, www.bflt.org/; Eastern Woodlands Rematriation Collective, Facebook, accessed April 12, 2023, www.facebook.com/EWRematriation/?ref=page_internal; Native American Land Conservancy, accessed April 12, 2023, www.nativeamericanland.org/; White Earth Land Recovery Project, accessed April 12, 2023, www.welrp.org/; "Native American Land Trusts and Conservation Groups," First Light, accessed April 12, 2023, https://firstlightlearningjourney.net/wp-content/uploads/2020/06/Roster-of-Native-American-Land-Trusts-2020-1.pdf.

22. "Distribution of the 10 Leading Causes of Death Among Black U.S. Residents in 2018," Statista, July 2021, www.statista.com/statistics/233310/distribution-of-the-10-leading-causes-of-death-among-african-americans/.

23. "Farm Subsidy Primer," EWG, accessed April 12, 2023, https://farm.ewg.org/subsidyprimer.php; Brad Plumer, "The $956 Billion Farm Bill in One Graph," *Washington Post*, January 28, 2014, www.washingtonpost.com/news/wonk/wp/2014/01/28/the-950-billion-farm-bill-in-one-chart/.

24. Roberto A. Ferdman, "The Disturbing Ways That Fast Food Chains Disproportionately Target Black Kids," *Washington Post*, November 12, 2014, www.washingtonpost.com/news/wonk/wp/2014/11/12/the-disturbing-ways-that-fast-food-chains-disproportionately-target-black-kids/.

25. ERS, "Food Security in the U.S.: Key Statistics & Graphics," USDA, last modified October 17, 2022, www.ers.usda.gov/topics/food-nutrition-assistance/food-security-in-the-u-s/key-statistics-graphics/; Alana Rhone et al., *Low-Income and Low-Supermarket-Access Census Tracts, 2010–2015* (Washington, DC: ERS, 2017), www.ers.usda.gov/webdocs/publications/82101/eib-165.pdf?v=3395.3.

26. "Black Communities Face Hunger at a Higher Rate Than Other Communities," Feeding America, accessed April 13, 2023, www.feedingamerica.org/hunger-in-america

/african-american; "Latino Hunger Facts," Feeding America, accessed April 13, 2023, www.feedingamerica.org/hunger-in-america/latino-hunger-facts; Sara Usha Maillacheruvu, "The Historical Determinants of Food Insecurity in Native Communities," Center on Budget and Policy Priorities, October 4, 2022, www.cbpp.org/research/food-assistance /the-historical-determinants-of-food-insecurity-in-native-communities#:~:text=From%20 2000%20to%202010%2C%2025,the%20rate%20of%20white%20Americans.

27. Nathaniel Meyersohn, "How the Rise of Supermarkets Left Out Black America," CNN Business, June 16, 2020, www.cnn.com/2020/06/16/business/grocery-stores-access -race-inequality/index.html.

28. *Good Food and Good Jobs for All*, Applied Research Center, July 10, 2012, www.race forward.org/research/reports/food-justice.

29. Jane E. Brody, "Prescribing Vegetables, Not Pills," *Well* (blog), *New York Times*, December 1, 2014, http://well.blogs.nytimes.com/2014/12/01/prescribing-vegetables-not-pills/.

30. Diane Pien, "Black Panther Party's Free Breakfast Program (1969–1980)," Black Past, February 11, 2010, www.blackpast.org/african-american-history/black-panther-partys-free -breakfast-program-1969-1980/.

31. Sweet Freedom Farm, accessed April 13, 2023, www.sweetfreedomfarm.org/; "Self-Determination Through Food Justice," Fresh Future Farm, accessed April 13, 2023, www .freshfuturefarm.org/; Rock Steady Farm, accessed April 13, 2023, www.rocksteadyfarm .com/; "Farm Share," Corbin Hill Food Project, accessed April 13, 2023, http://corbinhill -foodproject.org/farmshare.

32. Ed Whitfield, "Nevermind Guaranteed Income, We Want the Cow," Fund for Democratic Communities, January 2, 2017, https://f4dc.org/nevermind-guaranteed-income-we -want-the-cow/.

33. Cordula Droege et al., *The Right to a Remedy and Reparation for Gross Human Rights Violations*, rev. ed. (Geneva, Switzerland: International Commission of Jurists, 2018), www .icj.org/wp-content/uploads/2018/11/Universal-Right-to-a-Remedy-Publications-Reports -Practitioners-Guides-2018-ENG.pdf.

34. Jean Willoughby, "A Digital Map Leads to Reparations for Black and Indigenous Farmers," *Yes!*, February 21, 2018, www.yesmagazine.org/peace-justice/a-digital-map-leads -to-reparations-for-black-and-indigenous-farmers-20180221.

35. Henry Louis Gates Jr., "The Truth Behind 'Forty Acres and a Mule,'" *The African Americans: Many Rivers to Cross*, PBS, accessed April 13, 2023, www.pbs.org/wnet/african-americans -many-rivers-to-cross/history/the-truth-behind-40-acres-and-a-mule/; Dolores Barclay, Todd Lewan, and Allen G. Breed, "Prosperity Made Blacks a Target for Land Grabs," *Los Angeles Times*, December 9, 2001, www.latimes.com/archives/la-xpm-2001-dec-09-mn-13043-story .html; Abril Castro and Caius Z. Willingham, "Progressive Governance Can Turn the Tide for Black Farmers," Center for American Progress, April 3, 2019, www.americanprogress.org /issues/economy/reports/2019/04/03/467892/progressive-governance-can-turn-tide-black -farmers/; Vann R. Hewkirk II, "The Great Land Robbery," *Atlantic*, September 2019, www .theatlantic.com/magazine/archive/2019/09/this-land-was-our-land/594742/.

36. Melissa Gordon, "'Revolution Is Based on Land': Wealth Denied via Black Farmland Ownership Loss" (presentation, Tufts University, December 17, 2018), https://sites.tufts.edu/gis/files/2019/05/Gordon-Melissa_UEP232_Fall2018.pdf.

37. "Pay Up," *New York Times*, February 8, 2010, www.nytimes.com/2010/02/08/opinion/08mon3.html.

38. Land Loss Prevention Project, accessed April 13, 2023, www.landloss.org/.

39. Andrea Flynn et al., "Rewrite the Racial Rules: Building an Inclusive American Economy," Roosevelt Institute, June 6, 2016, http://rooseveltinstitute.org/rewrite-racial-rules-building-inclusive-american-economy/.

40. Karl Hamerschlag, "Fairness for Small Farmers: A Missing Ingredient in the U.S. Farm Bill," Fair World Project, March 6, 2013, http://fairworldproject.org/voices-of-fair-trade/fairness-for-small-farmers-a-missing-ingredient-in-the-u-s-farm-bill/.

41. Justice for Black Farmers Act, S. 300, 117th Cong. (2021), www.congress.gov/bill/117th-congress/senate-bill/300/text?q=%7B%22search%22%3A%5B%22justice+for+black+farmers+act%22%5D%7D&r=1&s=1.

42. Debbie Stabenow, "Senator Stabenow Announces the Urban Agriculture Act of 2016," news release, September 26, 2016, www.stabenow.senate.gov/news/senator-stabenow-announces-the-urban-agriculture-act-of-2016.

43. "Underserved and Veteran Farmers, Ranchers, and Foresters," USDA, accessed April 13, 2023, www.usda.gov/partnerships/underserved-veteran-farmers-ranchers-foresters.

44. Klaus W. Flach, Thomas O. Barnwell, and Pierre Crosson, "Impact of Agriculture on Atmospheric Carbon Dioxide," in *Soil Organic Matter in Temperate Agroecosystems: Long-Term Experiments in North America*, eds. Eldor A. Paul et al. (Boca Raton, FL: CRC Press, 1997).

45. Hannah Ritchie, Pablo Rosado, and Max Roser, "Environmental Impacts of Food Production," Our World in Data, 2022, https://ourworldindata.org/environmental-impacts-of-food.

46. Lisa Mahapatra, "The US Spends Less on Food Than Any Other Country in the World," *International Business Times*, January 23, 2014, www.ibtimes.com/us-spends-less-food-any-other-country-world-maps-1546945.

47. Shelby Vittek, "You Can Thank Black Horticulturalist Booker T. Whatley for Your CSA," *Smithsonian*, May 20, 2021, www.smithsonianmag.com/innovation/you-can-thank-black-horticulturist-booker-t-whatley-your-csa-180977771/; "1969: Fannie Lou Hamer Founds Freedom Farm Cooperative," SNCC Digital Gateway, accessed April 13, 2023, https://snccdigital.org/events/fannie-lou-hamer-founds-freedom-farm-cooperative/; New Communities, Inc., accessed April 13, 2023, www.newcommunitiesinc.com/new-communities.html#:~:text=To%20provide%20a%20safe%20haven,to%20become%20fully%20self%2Dsufficient.

48. Paul Hawken, ed., *Drawdown: The Most Comprehensive Plan Ever Proposed to Reverse Global Warming* (New York: Penguin, 2017).

49. "Conservation Agriculture," Project Drawdown, accessed April 13, 2023, https://drawdown.org/solutions/conservation-agriculture.

50. Natural Resources Conservation Service, "Environmental Quality Incentives Program," USDA, accessed April 13, 2023, www.nrcs.usda.gov/wps/portal/nrcs/main/national/programs/financial/eqip/.

51. "National Black Food Map & Directory," National Black Food and Justice Alliance, accessed April 13, 2023, www.blackfoodjustice.org/food-map-directory.

52. National Black Food and Justice Alliance, accessed April 13, 2023, www.blackfoodjustice.org/.

CHAPTER 6: THE ENVIRONMENTAL AND COMMUNITY IMPACTS OF INDUSTRIAL ANIMAL AGRICULTURE

1. "Quick Facts: Sussex County, Delaware," US Census Bureau, www.census.gov/quickfacts/sussexcountydelaware; USDA, "Table 19. Poultry—Inventory and Number Sold: 2017 and 2012," in *2017 Census of Agriculture—County Data*, www.nass.usda.gov/Publications/AgCensus/2017/Full_Report/Volume_1,_Chapter_2_County_Level/Delaware/st10_2_0019_0019.pdf.

2. "About Us," Mountaire, accessed April 13, 2023, https://mountaire.com/about-us/; Complaint, *Joseph Balback et al. v. Mountaire Farms of Delaware et al.*, S-18C-06-034 RFS (Del. Super., 2018).

3. Complaint, *Balback et al.*

4. DNREC, "Notice of Violation W-17-GWD-13," 2017, www.capegazette.com/sites/capegazette/files/2017/11/field/attachments/Mountaire%20NOV.PDF; Complaint, *Balback et al.*, 17; "Estimated Nitrate Concentrations in Groundwater Used for Drinking," US Environmental Protection Agency, last modified January 11, 2023, www.epa.gov/nutrient-policy-data/estimated-nitrate-concentrations-groundwater-used-drinking.

5. Complaint, *Balback et al.*; DNREC, "Notice of Violation."

6. Mary H. Ward et al., "Drinking Water Nitrate and Human Health: An Updated Review," *International Journal of Environmental Research and Public Health* 15, no. 7 (2018), https://doi.org/10.3390/ijerph15071557.

7. Ward et al., "Drinking Water Nitrate."

8. "Symptoms," E. Coli Homepage, CDC, last modified February 2, 2021, www.cdc.gov/ecoli/ecoli-symptoms.html.

9. "How We Got Here: A Brief History of Cattle and Beef Markets," *Feedlot*, last modified January 3, 2022, www.feedlotmagazine.com/news/industry_news/how-we-got-here-a-brief-history-of-cattle-and-beef-markets/article_1fe3c56c-6739-11ec-a5bb-2b72c548a8a4.html.

10. Carrie Hribar, *Understanding Concentrated Animal Feeding Operations and Their Impact on Communities* (Bowling Green, OH: National Association of Local Boards of Health, 2010), www.cdc.gov/nceh/ehs/docs/understanding_cafos_nalboh.pdf.

11. Tom Philpott, "A Reflection on the Lasting Legacy of 1970s USDA Secretary Earl Butz," *Grist*, February 8, 2008, https://grist.org/article/the-butz-stops-here/.

12. Mary K. Hendrickson et al., *The Food System: Concentration and Its Impacts* (Missouri: Farm Action, 2020).

13. "Per Capita Consumption of Poultry and Livestock, 1960 to Forecast 2023, in Pounds," National Chicken Council, last modified March 2023, www.nationalchickencouncil.org /statistic/per-capita-consumption-poultry/.

14. Mary Meisenzahl, "Chicken Prices Are Finally Falling and It's Triggering a Resurgence of Fast-Food Chicken Wars," *Business Insider*, November 29, 2022, www.businessinsider .com/popeyes-burger-king-wendys-mcdonalds-chicken-sandwich-wars-are-back-as-poultry -prices-fall-2022-11.

15. "Broiler Chicken Industry Key Facts 2022," National Chicken Council, www .nationalchickencouncil.org/statistic/broiler-industry-key-facts/; Lisa Held, "A Huge New Chicken CAFO in West Virginia Has Stoked Community Resistance," *Civil Eats*, April 7, 2021, https://civileats.com/2021/04/07/a-huge-new-chicken-cafo-in-west-virginia-has-stoked -community-resistance/.

16. Rockefeller Foundation, *True Cost of Food: Measuring What Matters to Transform the U.S. Food System*, July 2021, www.rockefellerfoundation.org/report/true-cost-of-food-measuring-what -matters-to-transform-the-u-s-food-system/.

17. Mark Eichmann, "Delaware Accuses Chicken Giant of Polluting Water," WHYY, November 10, 2017, https://whyy.org/articles/delaware-accuses-chicken-giant-polluting -water/.

18. Shawn M. Garvin, "Conciliation Order by Consent and Secretary's Order No. 2021-W-0013," DNREC, 2021, https://documents.dnrec.delaware.gov/Info/Documents/Secretarys -Order-No-2021-W-0013.pdf.

19. Complaint, *Balback et al.*

20. Kelli Steele, "Water Quality Monitoring Project Sheds New Light on Pollution Problem," Delaware Public Media, August 23, 2021, www.delawarepublic.org/science-health -tech/2021-08-23/water-quality-monitoring-project-sheds-new-light-on-pollution-problem; "Fish and Crab Kills in Indian River Caused by Nutrients/Heat," Delaware Center for the Inland Bays, August 31, 2018, www.inlandbays.org/projects-and-issues/issues/fish-and-crab -kills-in-indian-river-caused-by-nutrients-heat/.

21. Christina Cooke, "North Carolina's Factory Farms Produce 15,000 Olympic Pools Worth of Waste Each Year," *Civil Eats*, June 28, 2016, https://civileats.com/2016/06/28 /north-carolinas-cafos-produce-15000-olympic-size-pools-worth-of-waste/

22. Complaint, *Balback et al.*; DNREC, "Notice of Violation."

23. Keene Kelderman et al., *The Clean Water Act at 50: Promises Half Kept at the Half-Century Mark* (Washington, DC: Environmental Integrity Project, 2022), https:// environmentalintegrity.org/wp-content/uploads/2022/03/Revised-CWA-report-3.29.22.pdf.

24. Jayne Miller, "Report Finds Eastern Shore Chicken Farming a Main Cause of Chesapeake Bay Pollution," WBAL TV, October 28, 2021, www.wbaltv.com/article/report -eastern-shore-chicken-farming-chesapeake-bay-pollution/38080565.

25. "Historic Perdue Farmhouse in Salisbury, MD to Open for Tours," *Cape Gazette*, January 28, 2020, www.capegazette.com/article/historic-perdue-farmhouse-salisbury-md -open-tours/196247; Erica Shaffer, "Tyson Foods to Invest $300 Million Toward Production Plant in Virginia," *Meat + Poultry*, August 26, 2021, www.meatpoultry.com/articles/25425 -tyson-foods-to-invest-300-million-toward-production-plant-in-virginia.

26. "DCA Facts and Figures," Delmarva Chicken Association, www.dcachicken.com /facts/facts-figures.cfm.

27. Karl Blankenship, "Chesapeake Restoration Work Gets a Boost From Federal Funding," *Bay Journal*, March 31, 2022, www.bayjournal.com/news/pollution/chesapeake -restoration-work-gets-a-boost-from-federal-funding/article_27bbec58-af7b-11ec-bd0c -93eb4fed27f0.html.

28. Donnelle Eller, "50 Shades of Brown: Iowa Ranks No. 1 in, Ahem, No. 2, UI Researcher Calculates," *Des Moines Register*, June 10, 2019, www.desmoinesregister.com/story /money/agriculture/2019/06/10/iowa-leads-nation-poop-manure-university-iowa-livestock -clean-water-pollution-shades-brown-waste/1379973001/.

29. Lisa Held, "Inside the Rural Resistance to CAFOs," *Civil Eats*, March 3, 2020, https://civileats.com/2020/03/03/rural-resistance-builds-in-communities-facing-the-fallout -from-cheap-meat-production/.

30. Lisa Held, "Manure Overload from Factory Farms Spells Disaster for Waterways Across the Country," FoodPrint, June 15, 2020, https://foodprint.org/blog/manure-overload/.

31. Allison Kite, "Missouri Fines CAFO $18,000 for Polluting Streams with 300,000 Gallons of Waste," *Missouri Independent*, March 9, 2022, https://missouriindependent.com /2022/03/09/missouri-fines-cafo-18000-for-polluting-streams-with-300000-gallons-of-waste/.

32. "Bottom-Water Area of Hypoxia 1985–2022," What Is a Dead Zone?, National Ocean Service, National Oceanic and Atmospheric Administration, https://oceanservice .noaa.gov/facts/deadzone.html.

33. Complaint, *Balback et al.*

34. Madeleine Overturf, "DNREC and Mountaire Farms Enter Consent Decree After State Agency Files Complaints," WBOC, June 4, 2018, www.wboc.com/archive/dnrec -and-mountaire-farms-enter-consent-decree-after-state-agency-files-complaints/article _a4c45443-d9f5-5751-939e-7b849b9bafe5.html.

35. "Mountaire Offering Deep Well Installation to Millsboro Homes Affected by Wastewater Violation," WBOC, December 15, 2017, www.wboc.com/archive/mountaire -offering-deep-well-installation-to-millsboro-homes-affected-by-wastewater-violation/article _b0800e97-38c7-5f84-bc68-5b2dd3aba857.html.

36. Complaint, *Balback et al.*

37. Mallory Metzner, "Settlement Reached in One of Two Mountaire Farms Lawsuits," WBOC, February 20, 2020, www.wboc.com/archive/settlement-reached-in-one-of-two-mountaire-farms-lawsuits/article_fef5f958-f0e7-5244-8da0-721cf54eb135.html.

38. Held, "Rural Resistance."

39. Donnelle Eller, "For Safe Drinking Water, Des Moines Water Works Activates Rarely-Used Nitrate Removal Plant," *Des Moines Register*, June 9, 2022, www.desmoinesregister.com/story/money/agriculture/2022/06/09/des-moines-water-works-drinking-water-nitrate-removal-plant-activated-iowa/7570785001/.

40. "Kiwanis Burton Obituary," *Delaware State News*, March 13, 2014, www.legacy.com/us/obituaries/newszapde/name/kiwanis-burton-obituary?id=17772142.

41. Hribar, *Understanding Concentrated Animal Feeding Operations*.

42. Nina G. G. Domingo et al., "Air Quality–Related Health Damages of Food," *PNAS* 118, no. 20 (2021), https://doi.org/10.1073/pnas.2013637118.

43. Sara G. Rasmussen et al., "Proximity to Industrial Food Animal Production and Asthma Exacerbations in Pennsylvania, 2025–2012," *International Journal of Environmental Research and Public Health* 14, no. 4 (2017), https://doi.org/10.3390/ijerph14040362; Joan A. Casey et al., "Industrial Food Animal Production and Community Health," *Current Environmental Health Reports* 2 (2015), https://doi.org/10.1007/s40572-015-0061-0.

44. Cooke, "North Carolina's Factory Farms."

45. Melissa N. Poulsen et al., "High-Density Poultry Operations and Community-Acquired Pneumonia in Pennsylvania," *Environmental Epidemiology* 2, no. 2 (2018), https://doi.org/10.1097/EE9.0000000000000013.

46. Lisa Held, "The Battle Over Air Quality near Factory Farms on Maryland's Eastern Shore," *Civil Eats*, May 20, 2019, https://civileats.com/2019/05/20/the-battle-over-air-quality-near-factory-farms-on-marylands-eastern-shore/.

47. Richard Waite et al., "6 Pressing Questions About Beef and Climate Change, Answered," World Resources Institute, March 7, 2022, www.wri.org/insights/6-pressing-questions-about-beef-and-climate-change-answered.

48. Lisa Held, "The Field Report: New UN Climate Report Urges Food Systems Solutions—Before It's Too Late," *Civil Eats*, April 4, 2022, https://civileats.com/2022/04/04/the-field-report-new-un-climate-report-urges-food-systems-solutions-before-its-too-late/.

49. Gosia Wozniacka, "The Greenhouse Gas No One's Talking About: Nitrous Oxide on Farms, Explained," *Civil Eats*, September 19, 2019, https://civileats.com/2019/09/19/the-greenhouse-gas-no-ones-talking-about-nitrous-oxide-on-farms-explained/.

50. Lisa Held, "Methane from Agriculture Is a Big Problem. We Explain Why," *Civil Eats*, October 6, 2021, https://civileats.com/2021/10/06/methane-from-agriculture-is-a-big-problem-we-explain-why/.

51. Held, "Methane from Agriculture."

52. Nancy Matsumoto, "Is Grass-Fed Beef Really Better for the Planet? Here's the Science," NPR, August 13, 2019, www.npr.org/sections/thesalt/2019/08/13/746576239 /is-grass-fed-beef-really-better-for-the-planet-heres-the-science.

53. Josh Ulick, "Could Curbing Cow Burps Slow Global Warming?" *Wall Street Journal*, January 21, 2023, www.wsj.com/articles/could-curbing-cow-burps-slow-global-warming -11674267505.

54. Jennifer Hayden, "Cattle Are Part of the Climate Solution," Rodale Institute, August 28, 2020, https://rodaleinstitute.org/blog/cattle-are-part-of-the-climate-solution/.

55. Nassos Stylianou, Clara Guibourg, and Helen Briggs, "Climate Change Food Cal-culator: What's Your Diet's Carbon Footprint?," BBC News, August 9, 2019, www.bbc.com /news/science-environment-46459714.

56. Michael Grunwald, "What's the Most Climate-Friendly Way to Eat? It's Tricky," *Canary Media*, June 6, 2022, www.canarymedia.com/articles/food-and-farms/whats-the -most-climate-friendly-way-to-eat-its-tricky.

CHAPTER 7: A HISTORIC REVOLUTION IN THE FOOD SERVICE INDUSTRY

1. Bureau of Labor Statistics, *State Occupational Employment and Wage Estimates*, May 2019.

2. Saru Jayaraman, "What Created the American Crisis of Subminimum Pay?," *Lit-erary Hub*, November 5, 2021, https://lithub.com/what-created-the-american-crisis-of -subminimum-pay/.

3. Philip S. Foner and Ronald L. Lewis, eds., *The Black Worker*, vol. 1, *The Black Worker to 1896* (Philadelphia, PA: Temple University Press, 1978).

4. Kerry Segrave, *Tipping: An American Social History of Gratuities* (Jefferson, NC: McFarland, 1998).

5. Lawrence Tye, "Choosing Servility to Staff America's Trains," Alicia Patterson Foun-dation, June 9, 2003, https://aliciapatterson.org/lawrence-tye/choosing-servility-to-staff -americas-trains/.

6. One Fair Wage, *A Persistent Legacy of Slavery*, August 2020, https://onefairwage.site /persistent-legacy-of-slavery.

7. Rachel Tepper, "Lowest Paying Jobs in America: 7 Out of 10 Are in the Food Industry," *HuffPost*, April 2, 2013, www.huffpost.com/entry/lowest-paying-jobs-food-industry_n_2999799.

8. Lawrence Mishel, Elise Gould, and Josh Bivens, "Wage Stagnation in Nine Charts," Economic Policy Institute, January 6, 2015, www.epi.org/publication/charting-wage-stagnation/.

9. Mishel, Gould, and Bivens, "Wage Stagnation."

10. Sylvia Allegretto and David Cooper, "Twenty-Three Years and Still Waiting for Change," Economic Policy Institute, July 10, 2014, www.epi.org/publication/waiting-for-change-tipped -minimum-wage/.

11. One Fair Wage, *The Tipping Point: How the Subminimum Wage Keeps Income Low and Harassment High*, March 2021, https://onefairwage.site/wp-content/uploads/2021/03/OFW_TheTippingPoint_3-1.pdf.

12. One Fair Wage, *Take off Your Mask So I Know How Much to Tip You: Service Workers' Experience of Health & Harassment During COVID-19*, December 2020, https://onefairwage.site/wp-content/uploads/2020/12/OFW_COVID_WorkerExp-1.pdf.

13. One Fair Wage, *Intentional Inequality*, September 21, 2022, https://onefairwage.site/intentional-inequality.

14. National Restaurant Association, *State Statistics: 2017 and 2018 Estimated Sales, Eating and Drinking Establishments, and Eating and Drinking Places' Employees by State*, 2019. Estimated sales weighted by 2018 number of eating and drinking establishment employees.

15. "One Fair Wage: The Key to Saving the Restaurant Industry Post-COVID 19," One Fair Wage, accessed February 7, 2023, https://onefairwage.site/wp-content/uploads/2022/05/OFW_FactSheet_USA.pdf.

16. One Fair Wage, *Locked Out of Low Wages: Service Workers' Challenges with Accessing Unemployment Insurance During COVID-19*, May 2020.

17. One Fair Wage, *Take off Your Mask*.

18. Yea-Hung Chen et al., "Excess Mortality Associated with the COVID-19 Pandemic Among Californians 18–65 Years of Age, by Occupational Sector and Occupation: March Through October 2020," *PLoS One*, June 4, 2021, https://journals.plos.org/plosone/article?id=10.1371/journal.pone.0252454.

19. One Fair Wage, *Take off Your Mask*.

20. One Fair Wage, *The Great Black Restaurant Worker Exodus*, February 2022, https://onefairwage.site/the-great-black-restaurant-worker-exodus.

21. One Fair Wage analysis of US Bureau of Labor Statistics, "Quarterly Data 2019–2021, All Establishment Sizes," Quarterly Census of Employment and Wages, 2021.

22. One Fair Wage, *It's a Wage Shortage, Not a Worker Shortage*, May 2021, https://onefairwage.site/wage_shortage_not_a_worker_shortage.

23. One Fair Wage, *It's a Wage Shortage*.

24. One Fair Wage Employer Database, accessed February 7, 2023, https://docs.google.com/spreadsheets/d/1yigcr2ERmIOKorG83CUxDbdSNFK6ZvaggPoNdiK2VyQ/edit#gid=305801906.

25. One Fair Wage Employer Database.

26. One Fair Wage, *Raise the Wage Voter*, February 2021, https://raisethewagevoter.com/.

27. Lindsay VanHulle, "Michigan Legislature Passes Minimum Wage, Paid Sick Leave Bills to Avert Ballot," *Bridge Michigan*, September 5, 2018, www.bridgemi.com/michigan-government/michigan-legislature-passes-minimum-wage-paid-sick-leave-bills-avert-ballot.

28. "Washington, D.C., Initiative 82, Increase Minimum Wage for Tipped Employees Measure (2022)," Ballotpedia, https://ballotpedia.org/Washington,_D.C.,_Initiative_82,_Increase_Minimum_Wage_for_Tipped_Employees_Measure_(2022).

29. Paula Pecorella and Sam Quigley, "Restaurant Workers in D.C. Are Organizing to Overcome Corruption and Corporate Power," More Perfect Union, October 12, 2022, https://perfectunion.us/restaurant-workers-in-d-c-are-organizing-to-overcome-corruption-and-corporate-power/.

30. David A. Fahrenthold and Talmon Joseph Smith, "How Restaurant Workers Help Pay for Lobbying to Keep Their Wages Low," *New York Times*, January 17, 2023, www.nytimes.com/2023/01/17/us/politics/restaurant-workers-wages-lobbying.html.

CHAPTER 8: CHANGING THE CULTURE OF CAPITAL TO SUPPORT REGENERATIVE AGRICULTURE

1. Anil K. Giri et al., "Off-Farm Income a Major Component of Total Income for Most Farm Households in 2019," *Amber Waves* (blog), USDA, September 7, 2021, www.ers.usda.gov/amber-waves/2021/september/off-farm-income-a-major-component-of-total-income-for-most-farm-households-in-2019.

2. "Ag and Food Sectors and the Economy," ERS, USDA, last modified January 26, 2023, www.ers.usda.gov/data-products/ag-and-food-statistics-charting-the-essentials/ag-and-food-sectors-and-the-economy/?topicId=b7a1aba0-7059-4feb-a84c-b2fd1f0db6a3.

3. "Food Prices and Spending," ERS, USDA, last modified February 27, 2023, www.ers.usda.gov/data-products/ag-and-food-statistics-charting-the-essentials/food-prices-and-spending/.

4. "Leading Causes of Death," FastStats, CDC, last updated January 18, 2023, www.cdc.gov/nchs/fastats/leading-causes-of-death.htm.

5. Rockefeller Foundation, *True Cost of Food: Measuring What Matters to Transform the U.S. Food System*, July 2021, www.rockefellerfoundation.org/report/true-cost-of-food-measuring-what-matters-to-transform-the-u-s-food-system/.

6. "The Dust Bowl," National Drought Mitigation Center, accessed February 28, 2023, https://drought.unl.edu/dustbowl/.

7. "Assets, Debt, and Wealth," ERS, USDA, last modified February 7, 2023, www.ers.usda.gov/topics/farm-economy/farm-sector-income-finances/assets-debt-and-wealth/.

8. Brandi Janssen, "Safety Watch: Suicide Rate Among Farmers at Historic High," *Iowa Farmer Today*, December 10, 2016.

9. "Adoption of Genetically Engineered Crops in the U.S.," ERS, USDA, last modified September 14, 2022, www.ers.usda.gov/data-products/adoption-of-genetically-engineered-crops-in-the-u-s/.

10. Rick Smith, "Iowa's 'Black Gold' Is Washing Away," *Iowa Starting Line*, August 14, 2017, https://iowastartingline.com/2017/08/14/iowas-black-gold-washing-away/.

11. Sjoerd Willem Duiker and Jennifer Weld, "A Values and T Values: What Is That All About?," PennState Extension, last modified January 19, 2022, https://extension.psu.edu/a-values-and-t-values-what-is-that-all-about.

12. "Seed Commons' Approach to Non-Extractive Finance," Seed Commons, accessed February 28, 2023, https://seedcommons.org/about-seed-commons/seed-commons-approach-to-non-extractive-finance/.

13. "Farmland Ownership and Tenure," ERS, USDA, last modified May 16, 2022, www.ers.usda.gov/topics/farm-economy/land-use-land-value-tenure/farmland-ownership-and-tenure/.

14. Patrick Canning, "Where Do Americans' Food Dollars Go?," USDA, May 14, 2019, www.usda.gov/media/blog/2019/05/14/where-do-americans-food-dollars-go.

15. The farm safety-net programs include other components like agricultural disaster programs to offset the financial impact and production losses that result from natural disasters like flooding, fires, hurricanes, and drought. Assistance is available through various mechanisms like direct payments, loans, and cost shares to rehabilitate damaged land.

16. "Summary of Business," Risk Management Agency, USDA, accessed February 28, 2023, www.rma.usda.gov/SummaryOfBusiness.

17. "The United States Farm Subsidy Breakdown, 2020," Farm Subsidy Database, EWG, accessed February 28, 2023, https://farm.ewg.org/region.php?fips=00000&progcode=total&yr=2020.

18. "Farm Bill Primer: PLC and ARC Farm Support Programs," Congressional Research Service, May 16, 2022, https://crsreports.congress.gov/product/pdf/IF/IF12114.

19. Adam Andrzejewski, "Mapping the U.S. Farm Subsidy $1M Club," Forbes, August 14, 2018, www.forbes.com/sites/adamandrzejewski/2018/08/14/mapping-the-u-s-farm-subsidy-1-million-club/.

20. Anne Schechinger, " Under Trump, Farm Subsidies Soared and the Rich Got Richer," February 24, 2021, www.ewg.org/interactive-maps/2021-farm-subsidies-ballooned-under-trump/.

21. Andrzejewski, "Mapping."

22. Randy Schnepf, USDA's Actively Engaged in Farming (AEF) Requirement (Washington, DC: Congressional Research Service, 2016).

23. Ruixue Wang, Roderick M. Rejesus, and Serkan Aglasan, "Warming Temperatures, Yield Risk and Crop Insurance Participation," European Review of Agricultural Economics 48, no. 5 (December 2021): 1109–1131, https://doi.org/10.1093/erae/jbab034.

24. Bradley Zwilling, "Tenure Characteristics of Illinois Farmland," Farmdoc Daily 12, no. 92 (June 17, 2022), https://farmdocdaily.illinois.edu/2022/06/tenure-characteristics-of-illinois-farmland-4.html.

25. Neil D. Hamilton, "The Role of Land Tenure in the Future of American Agriculture," June 13, 2017, https://landforgood.org/wp-content/uploads/The-Role-of-Land-Tenure-by-Neil-Hamilton.pdf.

26. "Health and Economic Costs of Chronic Diseases," CDC, last modified March 23, 2023, www.cdc.gov/chronicdisease/about/costs/index.htm.

27. Cristin E. Kearns, Laura A. Schmidt, and Stanton A. Glantz, "Sugar Industry and Coronary Heart Disease Research: A Historical Analysis of Internal Industry Documents," *JAMA Internal Medicine* 176, no. 11 (November 2016): 1680, https://doi.org/10.1001/jamainternmed.2016.5394.

28. Olivier J. Wouters, "Lobbying Expenditures and Campaign Contributions by the Pharmaceutical and Health Product Industry in the United States, 1999–2018," *JAMA Internal Medicine* 180, no. 5 (March 3, 2020): 1–10, https://doi.org/10.1001/jamainternmed.2020.0146.

29. Donald R. Davis, Melvin D. Epp, and Hugh D. Riordan. "Changes in USDA Food Composition Data for 43 Garden Crops, 1950 to 1999," *Journal of the American College of Nutrition* 23, no. 6 (2004): 669–682, https://doi.org/10.1080/07315724.2004.10719409.

30. Louisa Burwood-Taylor, Rob Leclerc, and Robin Chauhan, *2022 AgFunder Agri-FoodTech Investment Report* (San Francisco: AgFunder, 2022).

31. Jim Monke, *Agricultural Credit: Institutions and Issues* (Washington, DC: Congressional Research Service, 2022), https://crsreports.congress.gov/product/pdf/R/R46768.

32. Chris Dall, "Report: US Pigs Consume Nearly as Many Antibiotics as People Do," Center for Infectious Disease Research and Policy, June 6, 2018, www.cidrap.umn.edu/antimicrobial-stewardship/report-us-pigs-consume-nearly-many-antibiotics-people-do.

33. Steve Suppan, "Is a Climate-Resilient, Racially-Just Farm Credit System Feasible?," Institute for Agriculture and Trade Policy, December 9, 2021, www.iatp.org/climate-resilient-racially-just-farm-credit-system.

34. "Soil Health Case Study Findings," American Farmland Trust, last modified September 2022, https://farmland.org/soil-health-case-studies-findings/; Maggie Monast, Laura Sands, and Alan Grafton, *Farm Finance and Conservation: How Stewardship Generates Value for Farmers, Lenders, Insurers and Landowners* (New York: Environmental Defense Fund, 2018), https://business.edf.org/files/farm-finance-report.pdf; Soil Health Partnership et al., *Conservation's Impact on the Farm Bottom Line* (New York: Environmental Defense Fund, 2021), https://business.edf.org/files/Conservation-Impact-On-Farm-Bottom-Line-2021.pdf.;

35. "Invest Your Values," As You Sow, accessed March 1, 2023, www.asyousow.org/invest-your-values.

36. "Transform Your Money," Money Transforms, accessed March 1, 2023, https://moneytransforms.com.

CHAPTER 10: POLITICS ON YOUR PLATE

1. "Food Deserts in the United States," Annie E. Casey Foundation, February 13, 2021, www.aecf.org/blog/exploring-americas-food-deserts.

2. Olivia Olson, "An UnHappy Meal: How Government Spending Forced Reliance on Fast Food," USC Bedrosian Center, December 18, 2018, https://bedrosian.usc.edu/an-unhappy-meal-how-government-spending-forced-reliance-on-fast-food/.

3. Georgina Gustin, "Big Meat and Dairy Companies Have Spent Millions Lobbying Against Climate Action, a New Study Finds," *Inside Climate News*, April 2, 2021, https://insideclimatenews.org/news/02042021/meat-dairy-lobby-climate-action/.

4. Fidel Ezeala-Harrison and John Baffoe-Bonnie, "Market Concentration in the Grocery Retail Industry: Application of the Basic Prisoners' Dilemma Model," *Advances in Management & Applied Economics* 6, no.1 (2016): 47–67, www.scienpress.com/Upload/AMAE/Vol%206_1_3.pdf.

5. Clare Kelloway and Sarah Miller, *Food and Power: Addressing Monopolization in America's Food System* (Washington, DC: Open Markets Institute, 2021), https://static1.squarespace.com/static/5e449c8c3ef68d752f3e70dc/t/614a2ebebf7d510debfd53f3/1632251583273/200921_MonopolyFoodReport_endnote_v3.pdf.

6. Melba Newsome, "Unchecked Growth of Industrial Animal Farms Spurs Long Fight for Environmental Justice in Eastern NC," *NC Health News*, October 20, 2021, www.northcarolinahealthnews.org/2021/10/20/environmental-justice-and-industrial-farming-in-eastern-nc/.

7. "Update: Exposing Fields of Filth: Factory Farms Disproportionately Threaten Black, Latino, and Native American North Carolinians," Waterkeeper Alliance, July 30, 2020, https://waterkeeper.org/news/update-exposing-fields-of-filth/.

8. Edward A. Chow et al., "The Disparate Impact of Diabetes on Racial/Ethnic Minority Populations," *Clinical Diabetes* 30, no. 3 (July 1, 2012), https://diabetesjournals.org/clinical/article/30/3/130/30876/The-Disparate-Impact-of-Diabetes-on-Racial-Ethnic.

9. "Policy Basics: Where Do Our Federal Tax Dollars Go?" Center on Budget and Policy Priorities, last modified July 28, 2022, www.cbpp.org/research/federal-budget/where-do-our-federal-tax-dollars-go.

10. Deboarh A. Galuska et al., "Addressing Childhood Obesity for Type 2 Diabetes Prevention: Challenges and Opportunities," *Diabetes Spectrum* 31, no. 4 (November 1, 2018), https://diabetesjournals.org/spectrum/article/31/4/330/32461/Addressing-Childhood-Obesity-for-Type-2-Diabetes.

11. Mark Bittman, "Telling Americans to 'Eat Better' Doesn't Work. We Must Make Healthier Food," *Guardian*, December 4, 2022, www.theguardian.com/commentisfree/2022/dec/04/americans-diet-public-health-food.

12. National Academies of Sciences, Engineering, and Medicine et al., *Redesigning the Process for Establishing the Dietary Guidelines for Americans: A Consensus Study Report of The National Academies of Science, Engineering, and Medicine* (Washington, DC: National Academies Press, 2017), https://nap.nationalacademies.org/catalog/24883/redesigning-the-process-for-establishing-the-dietary-guidelines-for-americans.

13. "Get the Facts: Added Sugars," CDC, last modified November 28, 2021, www.cdc.gov/nutrition/data-statistics/added-sugars.html.

14. Christopher R. Kelley, "An Overview of the Packers and Stockyards Act," *Arkansas Law Notes*, 2003, http://media.law.uark.edu/arklawnotes/files/2011/03/Kelley-An-Overview-of-the-Packers-and-Stockyards-Act-Arkansas-Law-Notes-2003.pdf.

15. "Packers and Stockyards Act," Agricultural Marketing Service, USDA, www.ams
.usda.gov/rules-regulations/packers-and-stockyards-act.

16. Cory Booker, "Booker, Tester, Merkley, Warren Introduce Bill to Impose Moratorium on Large Agribusiness Mergers," news release, May 18, 2022, www.booker.senate.gov
/news/press/booker-tester-merkley-warren-introduce-bill-to-impose-moratorium-on-large
-agribusiness-mergers.

17. Cory Booker, "Booker Reintroduces Bill to Reform Farm System with Expanded
Support from Farm, Labor, Environment, Pulic Health, Faith Based and Animal Welfare
Groups," news release, July 15, 2021, www.booker.senate.gov/news/press/booker-reintroduces
-bill-to-reform-farm-system-with-expanded-support-from-farm-labor-environment-public
-health-faith-based-and-animal-welfare-groups.

18. Randy Schnepf, *Farm-to-Food Price Dynamics* (Washington, DC: Congressional Research Service, 2015), https://sgp.fas.org/crs/misc/R40621.pdf; "Farm Share of U.S. Food
Dollar Reached Historic Low in 2021," ERS, USDA, last modified November 28, 2022,
www.ers.usda.gov/data-products/chart-gallery/gallery/chart-detail/?chartId=105281.

CHAPTER 11: THE FUTURE OF FOOD IS BLUE

1. "The Blue Food Assessment," Blue Food Assessment, accessed April 11, 2023, https://
bluefood.earth.

2. Margaret Cooney, Miriam Goldstein, and Emma Shapiro, "How Marine Protected
Areas Help Fisheries and Ocean Ecosystems," Center for American Progress, June 3, 2019,
www.americanprogress.org/article/marine-protected-areas-help-fisheries-ocean-ecosystems/.

3. Robert Jones, Bill Dewey, and Barton Seaver, "Aquaculture Critical for Feeding the
World in a Changing Climate," Nature Conservancy, January 25, 2022, www.nature.org
/en-us/what-we-do/our-insights/perspectives/aquaculture-food-world-changing-climate/.

4. Craig Dahlgren, *Review of the Benefits of No-Take Zones: A Report to the Wildlife Conservation Society* (New York: Wildlife Conservation Society, 2014).

5. "Explore: Why Lobsters and Scallops Are Larger and More Fertile in Lamlash Bay
No Take Zone," Community of Arran Seabed Trust, www.arrancoast.com; "2015 Goldman
Prize Winner Howard Wood," Goldman Environmental Prize, www.goldmanprize.org/recipient
/howard-wood/.

6. Kathy Gunst, "Seaweed Helps Maine Lobstermen Ride the Storm of Climate Change,"
Washington Post, July 25, 2022, www.washingtonpost.com/food/2022/07/25/seaweed-maine
-lobster-industry/.

7. Thomas Heaton, "Seaweed: A Healthy Boost to Pacific Island Economies?," *Honolulu
Civil Beat*, October 4, 2021, www.civilbeat.org/2021/10/seaweed-a-healthy-boost-to-pacific
-island-economies/.

8. Alex Brown, "Seaweed Farming Has Vast Potential (but Good Luck Getting a Permit)," *Stateline*, March 7, 2022, https://stateline.org/2022/03/07/seaweed-farming-has-vast
-potential-but-good-luck-getting-a-permit/.

9. *The State of the World's Fisheries and Aquaculture 2022* (Rome, Italy: UNFAO, 2022), www.fao.org/documents/card/en/c/cc0463en.

10. Degen Pener, "David E. Kelley on Starting a Fish Farm: 'I'm Trying to Save Something Bigger Than Myself,'" *Hollywood Reporter*, October 17, 2019, www.hollywoodreporter.com/news/general-news/david-e-kelley-staring-his-own-fish-farm-1248582/.

11. Carolyn Lange, "Shrimp Is a Jumbo Business in West Central Minnesota," *West Central Tribune*, October 24, 2021, www.wctrib.com/news/shrimp-is-a-jumbo-business-in-west-central-minnesota; Jeremy Laue, "Shrimp Farming in Minnesota? Yep, You Betcha," *Twin Cities Agenda*, June 23, 2017, https://tcagenda.com/2017/minnesota-shrimp-farming/.

CHAPTER 13: HEALTHY FOOD FOR ALL OUR KIDS

1. Allan S. Noonan, Hector Eduardo Velasco-Mondragon, and Fernando A. Wagner, "Improving the Health of African Americans in the USA: An Overdue Opportunity for Social Justice," *Public Health Reviews* 37, no. 12 (2016), https://publichealthreviews.biomedcentral.com/articles/10.1186/s40985-016-0025-4.

2. Edible Schoolyard NYC, accessed February 3, 2023, www.edibleschoolyardnyc.org/.

3. City of New York, "Mayor Adams and Chancellor Banks Announce Launch of Inaugural Chefs Council," news release, September 27, 2022, www.nyc.gov/office-of-the-mayor/news/703-22/mayor-adams-chancellor-banks-launch-inaugural-chefs-council.

4. "ScratchWorks: The Collective Power of Good Food in Schools," ScratchWorks, accessed February 3, 2023, https://wearescratchworks.org/.

CHAPTER 14: THE FOUR BITES

1. "Meat, Poultry and Seafood Market Worth $1,601.2 Billion by 2030," Grand View Research, August 2022, www.grandviewresearch.com/press-release/global-meat-poultry-seafood-market.

2. *Tackling Climate Change Through Livestock: A Global Assessment of Emissions and Mitigation Opportunities* (Rome, Italy: UNFAO, 2013).

3. Anthony D. Barnosky, "Megafauna Biomass Tradeoff as a Driver of Quaternary and Future Extinctions," *PNAS* 105 (August 12, 2008), https://doi.org/10.1073/pnas.0801918105.

4. Meagan E. Schipanski and Elena M. Bennett, "The Influence of Agricultural Trade and Livestock Production on the Global Phosphorus Cycle," *Ecosystems* 15, no. 2 (March 2012): 256–268, www.jstor.org/stable/41413041.

5. "Global Meat Consumption, World, 1961 to 2050," Our World in Data, accessed March 19, 2023, https://ourworldindata.org/grapher/global-meat-projections-to-2050.

6. Brian Machovina, Kenneth J. Feeley, and William J. Ripple, "Biodiversity Conservation: The Key Is Reducing Meat Consumption," *Science of the Total Environment* 536 (December 1, 2015), www.sciencedirect.com/science/article/abs/pii/S0048969715303697.

7.   David Tilman and Michael Clark, "Global Diets Link Environmental Sustainability and Human Health," *Nature* 515 (2014), www.nature.com/articles/nature13959.

8.   Mark Bittman, "What's Wrong with What We Eat," TED video, 2007, accessed March 30, 2023, www.ted.com/talks/mark_bittman_what_s_wrong_with_what_we_eat /transcript?language=en.

9.   Allan Savory, "How to Fight Desertification and Reverse Climate Change," filmed at TED2013 conference, TED video, www.ted.com/talks/allan_savory_how_to_fight _desertification_and_reverse_climate_change.

10. For more on competing paradigms in food see Warren Belasco, *Meals to Come: A History of the Future of Food* (Berkeley: University of California Press, 2006); Anthony J. McMichael, "Globalization, Climate Change, and Human Health," *New England Journal of Medicine* (2013), www.nejm.org/doi/full/10.1056/NEJMra1109341; Ledivow et al., 2012; Chiles, 2013; and Tim Lang and Michael Heasman, *Food Wars: The Global Battle for Mouths, Minds and Markets*, 2nd ed. (London: Routledge, 2016).

11. Cameron Bruett, "Cameron Bruett—Our Shared Journey of Continuous Improvement," Truffle Media, YouTube video, May 15, 2015, www.youtube.com/watch?v =Vu8F4isWDJE.

12. Cameron Bruett, "Defining Sustainability for the Cattle Industry | Cameron Bruett, JBS USA | TCC 2015," Noble Research Institute, YouTube video, March 21, 2015, www .youtube.com/watch?v=P1dxuX4glqc.

13. "When We're Dead and Buried, Our Bones Will Keep Hurting," Human Rights Watch, September 4, 2019, www.hrw.org/report/2019/09/04/when-were-dead-and-buried-our -bones-will-keep-hurting/workers-rights-under-threat.

14. Jonathan A. Foley et al., "Solutions for a Cultivated Planet," *Nature* 478 (October 12, 2011), www.nature.com/articles/nature10452.

15. "Smart Agriculture Market Size, Share & Trends Analysis Report by Type," Grand View Research, accessed March 28, 2023, www.grandviewresearch.com/industry-analysis /smart-agriculture-farming-market.

16. Alessandro Bonanno and Douglas H. Constance, "Globalization, Fordism, and Post-Fordism in Agriculture and Food: A Critical Review of the Literature," *Culture & Agriculture* 23 (2008), https://doi.org/10.1525/cag.2001.23.2.1.

17. "Sustainability," JBS Foods, https://jbsfoodsgroup.com/our-purpose/sustainability.

18. Naira Hofmeister et al., "JBS Admits to Buying Almost 9,000 Cattle from 'One of Brazil's Biggest Deforesters,'" *Unearthed*, November 11, 2022, https://unearthed.greenpeace .org/2022/11/11/jbs-cattle-brazils-biggest-deforester-amazon/.

19. *Forks over Knives*, directed by Lee Fulkerson (2011), www.forksoverknives.com/the-film/.

20. Suzy Amis Cameron, *OMD: The Simple, Plant-Based Program to Save Your Health, Save Your Waistline, and Save the Planet* (New York: Atria, 2018).

21. Hannah Ritchie, "If the World Adopted a Plant-Based Diet We Would Reduce Global Agricultural Land Use from 4 to 1 Billion Hectares," Our World in Data, March 4, 2021, https://ourworldindata.org/land-use-diets.

22. Rob Bailey, Antony Froggatt, and Laura Wellesley, *Livestock—Climate Change's Forgotten Sector* (London: Chatham House, 2014).

23. T. Colin Campbell and Thomas M. Campbell II, *The China Study: The Most Comprehensive Study of Nutrition Ever Conducted and the Startling Implications for Diet, Weight Loss, and Long-Term Health* (Dallas, TX: BenBella Books, 2006).

24. John A. McDougall, *The Starch Solution: Eat the Foods You Love, Regain Your Health, and Lose the Weight for Good!* (New York: Rodale, 2013).

25. "Undo It with Ornish," Ornish Lifestyle Medicine, www.ornish.com/undo-it/.

26. Note that the Blue Zones populations' diets varied widely, and none were entirely vegan.

27. Dan Buettner, *The Blue Zones: Lessons for Living Longer from the People Who've Lived the Longest* (Washington, DC: National Geographic Books, 2010).

28. Nils-Gerrit Wunsch, "Veganism and Vegetarianism Worldwide—Statistics & Facts," Statista, December 9, 2021, www.statista.com/topics/8771/veganism-and-vegetarianism -worldwide/#topicOverview.

29. Barry M. Popkin, "Part II. What Is Unique About the Experience in Lower and Middle-Income Less-Industrialised Countries Compared with the Very-High-Income Industrialised Countries?," *Public Health Nutrition* (2002), https://doi.org/10.1079/PHN2001295.

30. Commodity Promotion, Research, and Information Act of 1996, 7 U.S.C. 7401 (1996), accessed March 28, 2023, https://uscode.house.gov/view.xhtml?req=(title:7%20section: 7401%20edition:prelim).

31. Tim Lang, David Barling, and Martin Caraher, *Food Policy: Integrating Health, Environment and Society* (Oxford: Oxford University Press, 2009).

32. Mark Hyman, *The Pegan Diet: 21 Practical Principles for Reclaiming Your Health in a Nutritionally Confusing World* (New York: Little, Brown Spark, 2021).

33. Life-cycle assessments (or analysis) measure the "cradle to grave" footprints of a product or food—that is, inputs from origin through production including energy use, transportation, processing, and end use. Some solve for energy or carbon, others water or other environmental metrics.

34. Ulf Sonesson et al., "Protein Quality as Functional Unit—A Methodological Framework for Inclusion in Life Cycle Assessment of Food," *Journal of Cleaner Production* 140, part 2 (2017), www.sciencedirect.com/science/article/abs/pii/S0959652616307946.

35. Patrizia Cook, recording from an interview, Maldonado, Uruguay, September 19, 2019.

36. "Overview of Greenhouse Gases," Environmental Protection Agency, last modified April 13, 2023, www.epa.gov/ghgemissions/overview-greenhouse-gases.

37. Daniela Ibarra-Howell, in-person interview on regenerative beef, Boulder, CO, 2020.

38. Lautaro Pérez Rocha, in-person interview at a Savory Regenerative Beef gathering, Maldonado, Uruguay, September 19, 2019.

39. "Ecological Outcome Verification," Savory Institute, https://savory.global/eov/.

40. Mariko Thorbecke and Jon Dettling, "Carbon Footprint Evaluation of Regenerative Grazing at White Oak Pastures" (presentation, Quantis, February 25, 2019), https://blog.whiteoakpastures.com/hubfs/WOP-LCA-Quantis-2019.pdf?_hsenc=p2ANqtz-9XE6 zsk6P7ux9IybulX8CBvXiFNzlplqmy5xfZLLwcUjCUwpgkoaNsv_3On6uddObaLtMemK pUfLTFxDCkDxwdk2mB3XrK12FJxO-BxYoRei_4Yn8.

41. "Managed Grazing," Project Drawdown, accessed March 27, 2023, https://drawdown .org/solutions/managed-grazing.

42. "For the Sake of Sustainability: Ted Turner Makes an Urgent Appeal to Reverse Population Growth," Bridgespan Group, November 27, 2013, www.bridgespan.org/insights /ted-turner/for-the-sake-of-sustainability-ted-turner-makes-a.

43. William Godwin, *Of Population: An Enquiry Concerning the Power of Increase in the Numbers of Mankind* (London: Longman, Hurst, Rees, Orme and Brown, 1820), 488, 498.

44. Belasco, *Meals to Come.*

45. Belasco, *Meals to Come.*

46. Kate O'Riordan, Aristea Fotopoulou, and Neil Stephens, "The First Bite: Imaginaries, Promotional Publics and the Laboratory Grown Burger," *Public Understanding of Science* 26, no. 2 (2017), https://doi.org/10.1177/0963662516639001.

47. Erik Jönsson, "Benevolent Technotopias and Hitherto Unimaginable Meats: Tracing the Promises of In Vitro Meat," *Social Studies of Science* 46 (2016), https://doi.org/10.1177 /0306312716658561.

48. *Cultivated Meat and Seafood: 2021 State of the Industry Report,* Good Food Institute, 2022, https://gfi.org/wp-content/uploads/2022/04/2021-Cultivated-Meat-State-of-the-Industry -Report-1.pdf.

49. Uma Valeti, "Uma Valeti from Memphis Meats at EAT Stockholm Food Forum 2017," EAT, YouTube video, June 13, 2017, https://youtu.be/S2m_YtqkGGk.

50. Mark Post, "Meet the New Meat," filmed at TEDxHaarlem, TEDx Talks, YouTube video, June 20, 2016, https://youtu.be/ZExbQ8dkJvc.

51. *Plant-Based Meat Market Size, Share & Trends Analysis Report by Source, by Product, by Type,* report no. GVR-4-68039-145-9, Grand View Research, 2023.

52. Ethan Brown, "Ethan Brown - Beyond Meat," ideacity, YouTube video, February 14, 2017, www.youtube.com/watch?v=4x8jfiaLCPY&ab_channel=ideacity.

53. Gregory A. Keoleian and Martin C. Heller, *Beyond Meat's Beyond Burger Life Cycle Assessment: A Detailed Comparison Between a Plant-Based and an Animal-Based Protein Source* (Ann Arbor: Center for Sustainable Systems, 2018), 1–38.

54. Hanna L. Tuomisto and M. Joost Teixeira de Mattos, "Environmental Impacts of Cultured Meat Production," *Environmental Science and Technology* (2011), https://pubs.acs.org/doi/10.1021/es200130u.

55. "Food Safety," WHO, accessed March 30, 2023, www.who.int/news-room/fact-sheets/detail/food-safety.

56. "Reducing the Price of Alternative Proteins," Good Food Institute, accessed April 7, 2023, https://gfi.org/wp-content/uploads/2021/12/Reducing-the-price-of-alternative-proteins_GFI_2022.pdf.

57. Ashkan Pakseresht, Sina Ahmadi Kaliji, and Maurizio Canavari, "Review of Factors Affecting Consumer Acceptance of Cultured Meat," *Appetite* 170 (March 1, 2022), https://doi.org/10.1016/j.appet.2021.105829.

58. Charles Eisenstein, *The Yoga of Eating: Transcending Diets and Dogma to Nourish the Natural Self* (Brandywine, MD: NewTrends Publishing, Inc., 2003).

59. Language inspired by Charles Eisenstein's book title *The More Beautiful World Our Hearts Know Is Possible* (Berkeley, CA: North Atlantic Books, 2013).

INDEX

Western diet, 9

Whatley, Booker T., 85, 86

White, Noel, 21

White Earth Land Recovery Project, 79

White House Conference on Hunger, Nutrition, and Health, 155–157

White Oak Pastures (WOP), 224, 230

Whitfield, Ed, 82

whole foods, 30, 37, 43, 180

Whole Foods (supermarket chain), 8, 141, 149

Whole Kids Foundation, 199

Wholesome Wave, 267

WIC. *See* Special Supplemental Nutrition Program for Women, Infants, and Children

wildcat strikes, 8

Williams, Billy, 13

Wilson, Matte, 260

Wisconsin, 58–60, 64, 68, 70, 266

Wisconsin Farmers Union, 69, 71

Wisconsin Food Hub Cooperative, 57

Wise, Timothy A., 66

WITS. *See* Wellness in the Schools

women, 106, 258–259

Wood, Howard, 166, 167

WOP. *See* White Oak Pastures

worker-driven social responsibility (WSR), 140, 144–145

workers

essential, 7, 14

farmworker justice, 76–77

guest, 8, 143

immigrant, 24

injuries, 16–20

in meat production, 210–211

supporting, 262–263

World Health Organization, 34, 47, 156

World War II, 18, 63–64

World Wildlife Fund (WWF), 211, 269

WSR. *See* worker-driven social responsibility

WWF. *See* World Wildlife Fund

xenobiotics, 30

Yale School of Management, 251

youth, recognizing and supporting, 261–262

YouTube, 205

YPARD, 261

Zimbabwe, 259

Zimberoff, Larissa, 177

Zimmern, Andrew, 161–163

I believe that a good story well told can truly make a difference in how one sees the world. This is why I started **Participant** to tell compelling, entertaining stories that create awareness of the real issues that shape our lives.

At Participant, we seek to entertain our audiences first, and then invite them to participate in making a difference. With each film, we create social action and advocacy programs that highlight the issues that resonate in the film and provide ways to transform the impact of the media experience into individual and community action.

Over 100 films later, from GOOD NIGHT AND GOOD LUCK, to AN INCONVENIENT TRUTH, and **from WONDER to ROMA**, and through thousands of social action activities, Participant continues to create entertainment that inspires and compels social change. Now through our partnership with PublicAffairs, we are extending our mission so that more of you can join us in making our world a better place.

Jeff Skoll
Founder